# Methods in Molecular Biology

Volume 2

# NUCLEIC ACIDS

## Biological Methods

***Methods in Molecular Biology,*** edited by **John M. Walker,** *1984*
Volume I: ***Proteins***
Volume II: ***Nucleic Acids***

***Liquid Chromatography in Clinical Analysis,*** edited by **Pokar M. Kabra** and **Laurence J. Marton,** *1981*

***Metal Carcinogenesis Testing:*** *Principles and In Vitro Methods,* by **Max Costa,** *1980*

# Methods in Molecular Biology

Volume 2

# NUCLEIC ACIDS

Edited by

**John M. Walker**

**Humana Press • Clifton, New Jersey**

**Library of Congress Cataloging in Publication Data**
(Revised for volume 2)

Main entry under title

Methods in molecular biology.

(Biological methods)
Includes bibliographies and indexes.
Contents: v. 1. Proteins—v. 2. Nucleic acids.
1. Molecular biology—Technique—Collected works.
I. Walker, John M., 1948– II. Series.
QH506.M45 1984 574.8'8'078 84-15696
ISBN 0-89603-062-8 (v. 1)
ISBN 0-89603-064-4 (v. 2)


Crescent Manor
PO Box 2148
Clifton, NJ 07015



Printed in the United States of America

# Preface

In recent years there has been a tremendous increase in our understanding of the functioning of the cell at the molecular level. This has been achieved in the main by the invention and development of new methodology, particularly in that area generally referred to as "genetic engineering". While this revolution has been taking place in the field of nucleic acids research, the protein chemist has at the same time developed fresh methodology to keep pace with the requirements of present day molecular biology. Today's molecular biologist can no longer be content with being an expert in one particular area alone. He/she needs to be equally competent in the laboratory at handling DNA, RNA, and proteins, moving from one area to another as required by the problem he/she is trying to solve. Although many of the new techniques in molecular biology are relatively easy to master, it is often difficult for a researcher to obtain all the relevant information necessary for setting up and successfully applying a new technique. Information is of course available in the research literature, but this often lacks the depth of description that the new user requires. This requirement for in-depth practical details has become apparent by the considerable demand for places on our Molecular Biology Workshops held at Hatfield each summer. This book is therefore an attempt to provide detailed protocols for many of the basic techniques necessary for working with DNA, RNA, and proteins. This volume gives practical procedures for a wide range of nucleic acid techniques. A companion volume (Volume 1) provides coverage for protein techniques. Each method is described by an author who has regularly used the technique in his or her own laboratory. Not all the techniques described necessarily represent the state-of-the-art. They are, however, dependable methods that achieve the desired result.

Each chapter starts with a description of the basic theory behind the method being described. However, the main aim of this book is to describe the practical steps necessary for carrying out the method successfully. The Methods section therefore contains a detailed step-by-step description of a protocol that will result in the successful execution of the method. The Notes section complements the Methods section by indicating any major problems or faults that can occur with the technique, and any possible modifications or alterations.

This book should be particularly useful to those with no previous experience of a technique, and, as such, should appeal to undergraduates (especially project students), postgraduates, and research workers who wish to try a technique for the first time.

**John M. Walker**

# Contents

## Contributors

ALAN BATESON • Department of Biochemistry, Queen Elizabeth College, University of London, London, England

STEPHEN A. BOFFEY • Division of Biological and Environmental Sciences, The Hatfield Polytechnic, Hatfield, Hertfordshire, England

C. D. BOYD • MRC Unit for Molecular and Cellular Cardiology, University of Stellenbosch Medical School, Tygerberg, South Africa

KOKILA CHOTAI • Department of Biochemistry, Queen Elizabeth College, University of London, London, England

CLAUS CHRISTIANSEN • Gensplejsningsgruppen, The Technical University of Denmark, Lyngby, Denmark

CRAIG A. COONEY • Department of Biological Chemistry, School of Medicine, University of California, Davis, California

J. W. DALE • Department of Microbiology, University of Surrey, Guildford, Surrey, England

WIM GAASTRA • Department of Microbiology, The Technical University of Denmark, Lyngby, Denmark

ELLIOT B. GINGOLD • Division of Biological and Environmental Sciences, The Hatfield Polytechnic, Hatfield, Hertfordshire, England

P. J. GREENAWAY • Molecular Genetics Laboratory, PHLS Centre for Applied Microbiology and Research, Porton Down, Salisbury, Wilts., United Kingdom

A. GROBLER-RABIE • MRC Unit for Molecular and Cellular Cardiology, University of Stellenbosch Medical School, Tygerberg, South Africa

ELIZABETH G. GURNEY • Department of Biology, University of Utah, Salt Lake City, Utah

THEODORE GURNEY, JR. • Department of Biology, University of Utah, Salt Lake City, Utah

KIRSTEN HANSEN • Department of Microbiology, The Technical University of Denmark, Lyngby, Denmark

VERENA D. HUEBNER • University of California, Department of Biological Chemistry, School of Medicine, Davis, California

JYTTE JOSEPHSEN • Department of Microbiology, The Technical University of Denmark, Lyngby, Denmark

PER LINÅ JØRGENSEN • Department of Microbiology, The Technical University of Denmark, Lyngby, Denmark

PER KLEMM • Department of Microbiology, The Technical University of Denmark, Lyngby, Denmark

YUNUS LUQMANI • Department of Biochemistry, Queen Elizabeth College, University of London, London, United Kingdom

C. G. P. MATHEW • MRC Molecular and Cellular Cardiology Research Unit, University of Stellenbosch Medical School, Tygerberg, South Africa.

HARRY R. MATTHEWS • University of California, Department of Biological Chemistry, School of Medicine, Davis, California

R. McGOOKIN • Inveresk Research International Limited, Musselburgh, Scotland

C. L. OLLIVER • MRC Unit for Molecular and Cellular Cardiology, University of Stellenbosch Medical School, Tygerberg, South Africa

M. J. OWEN • Department of Zoology, University College London, London, England

JEFFREY W. POLLARD • Department of Biochemistry, Queen Elizabeth College, University of London, London, United Kingdom

ROBERT J. SLATER • Division of Biological and Environmental Sciences, The Hatfield Polytechnic, Hatfield, Hertfordshire, England

JAAP H. WATERBORG • Department of Biological Chemistry, University of California School of Medicine, Davis, California.

# Chapter 1

# The Burton Assay for DNA


***Jaap H. Waterborg and Harry R. Matthews***

*Department of Biological Chemistry, University of California School of Medicine, Davis, California*


## Introduction

The Burton assay for DNA is a colorimetric procedure for measuring the deoxyribose moiety of DNA. It is reasonably specific for deoxyribose, although very high concentrations of ribose (from RNA) or sucrose must be avoided. The method can be used on relatively crude extracts and in other circumstances where direct measurement of ultraviolet absorbance of denatured DNA is not practical. The assay has been widely used.

## Materials

1. Diphenylamine reagent: Dissolve 1.5 g of diphenylamine in 100 mL glacial acetic acid. Add 1.5 mL of concentrated (98–100%) $H_2SO_4$ and mix well. Store this reagent in the dark. Just before use, add 0.5 mL of acetaldehyde stock solution.

2. Acetaldehyde stock: 2 mL acetaldehyde in 100 mL distilled water. Store at 4°C where it is stable for a few months.
3. 1*N* perchloric acid (PCA).
4. *Standards*: Dilute a DNA stock solution with distilled water as follows:

| | | | | | | |
|---|---|---|---|---|---|---|
| DNA stock, μL | 0 | 10 | 20 | 50 | 100 | 200 |
| Water, mL | 1.0 | 0.990 | 0.980 | 0.950 | 0.900 | 0.800 |
| DNA concentration, μg/mL | 0 | 10 | 20 | 50 | 100 | 200 |

5. DNA stock: 1 mg/mL in distilled water. Store frozen at −20°C where it is stable for a few months.

## Method

1. Extract the sample as required (*see* Notes).
2. Add 0.5 mL of 1*N* PCA to 0.5 mL of sample or standard. Hydrolyze for 70 min at 70°C.
3. Cool the hydrolyzed samples on ice for 5 min. Centrifuge (1500*g*; 5 min; 4°C) and decant the supernatants into marked tubes.
4. Add 1 mL of 0.5*N* PCA to each pellet, vortex, repeat step 3, and carry the combined supernatants forward to step 5. (This step is optional; *see* Note 3)
5. Add 2 vol. of diphenylamine reagent to 1 vol of the supernatants (0.5*N* PCA hydrolyzates from step 3). Mix and incubate at 30°C for 18 hr.
6. Read the absorbance at both 595 and 650 nm, using the 0 μg/mL standard as a blank.
7. Plot a standard curve of absorbance at 595 nm minus absorbance at 650 nm as a function of initial DNA concentration and then use the curve to read off unknown DNA concentrations.

## Notes

1. Extraction conditions may have to be optimized for particular applications. The following procedures are routinely used in our laboratory.

(a) This extraction is required if the sample contains mercaptoethanol, dithiothreitol, or other interfering low molecular weight substances. It is also required as a preliminary if extraction (b), which is for whole cells or organelles that contain lipids, is to be used. Add 1 vol. of 0.2*N* PCA in 50% ethanol:50% distilled water and mix by vortexing. Cool on ice for 15 min and then centrifuge (5 min, 1500*g*, 4°C). Discard the supernatant.

(b) To the pellet add 1 mL of ethanol–ether (3:1, v/v). Incubate for 10 min at 70°C. Centrifuge (5 min, 1500*g*) and discard the supernatant. To the pellet add 1 mL of ethanol (96%), vortex and centrifuge (5 min, 1500*g*). Discard the supernatant.

2. If the sample is a pellet, at step 2 add 1 mL of 0.5*N* PCA to the pellet and proceed with the hydrolysis at 70°C. If the sample is too dilute ( $< 10$ μg/mL) and is available in a volume larger than 0.5 mL, then it may be concentrated by precipitation, as described in Note 1, extraction (a), or by precipitation with 0.5*N* PCA.
3. Step 4 is optional. It provides a more quantitative recovery of nucleic acid in the supernatant, but reduces the sensitivity of the overall assay.
4. The diphenylamine reagent is not water soluble. Rinse out glassware with ethanol before washing in water. Take care to use a dry spectrophotometer cuvet and clean it with ethanol.
5. It is recommended to run a standard curve with each group of assays, preferably in duplicate. Duplicate or triplicate unknowns are recommended.

## References

1. Burton, K. (1956) A study of the conditions and mechanism of the diphenylamine reaction for the colorimetric estimation of DNA. *Biochem. J.* **62,** 315–323.

# Chapter 2

# DABA Fluorescence Assay for Submicrogram Amounts of DNA

*Theodore Gurney, Jr. and Elizabeth G. Gurney*

*Department of Biology, University of Utah, Salt Lake City, Utah*

## Introduction

The fluorescence assay of Kissane and Robins (*1*) is used to quantify deoxypurine nucleosides in crude mixtures. Acid-catalyzed depurination exposes the 1′ and 2′ carbons of deoxyribose, which can then form a strongly fluorescent compound with diaminobenzoic acid (DABA). DABA can react with all aldehydes of the form $RCH_2CHO$, but deoxyribose is the predominant one in mammalian cells and essentially the only one in the acid or alcohol precipitates of aqueous extracts. Hence, no purification is required and RNA does not interfere. In our hands, the method is useful down to 30 ng of DNA, and probably could be made more sensitive, as discussed below. The method requires a visible-light fluorometer; the excitation wavelength is near 410 nm, with maximum fluorescence near 510 nm (2).

## Materials

1. The purity of DABA (3,5-diaminobenzoic acid dihydrochloride) determines the background of the assay, and hence the sensitivity. Purified DABA, in dihydrochloride form, can be purchased, or else crude commercial material can be purified. DABA should be white with a slight yellow-green fluorescence. Brown or grey powder usually gives high background. A procedure for purifying DABA is given below. DABA is stored in powder form at −20°C and is stable for years.
2. Highly purified mammalian DNA or salmon sperm DNA is used to calibrate the assay. DNA is dissolved in water at concentrations of 50, 100, and 300 μg/mL and is stored in quantities of 1 mL at −20°C. One set of the dilutions suffices for at least 20 calibration curves. Concentrations of DNA are measured in 0.1*M* NaCl assuming $E^{1\%}_{260} = 200$, i.e., 10 μg gives $A_{260} = 0.2$.
3. Purified RNA is used to coprecipitate DNA from dilute solutions. Commerical yeast RNA, free from detectable DNA, can be stored in powder form at −20°C. RNA is used in a precipitation buffer, described below.
4. A fluorometer or a spectrofluorometer. The excitation wavelength is 410 nm and the emission wavelength is 510 nm. Hinegardner (2) describes specific equipment.
5. Precipitation buffer: 100 m*M* NaCl; 10 m*M* Tris HCl, pH 7.5; 1 m*M* EDTA; 100 μg/mL yeast RNA. The buffer is stored at +2°C, or at −20°C if it is not used frequently. Bacterial growths can certainly raise the background.
6. Trichloroacetic acid is prepared as a 20% (w/v) solution and stored at +2°C.

## Methods

### *Purification of DABA*

This procedure, based on that of Hinegardner (2), will yield better DABA than the best commercial material, starting with inexpensive crude DABA. Yields should be 50–70%.

1. Put 100 g of crude (dark brown) DABA in a 1 L beaker, add 250 mL of distilled water, and stir to dissolve at room temperature in a fume hood.
2. Add 250 mL of concentrated HCl and stir slowly with a glass rod. A precipitate will form. Collect the precipitate on Whatman #1 paper using a Buchner funnel and suction. The suspension is thixotropic, so it is necessary to shake the funnel while filtering.
3. Redissolve the precipitate in 250 mL of water, or more if necessary, and then add an equal volume (to that of added water) of concentrated HCl.
4. Filter as above. If necessary, repeat the solution–precipitation cycles until the color of the precipitate is no darker than light brown.
5. Dissolve the precipitate in the minimum amount of water necessary. Measure the water volume, and then add 15 mg of activated charcoal powder "Norit A" per mL of added water. Stir to make a uniform suspension and then let the suspension rest unstirred for 30 min.
6. Centrifuge the suspension (5000*g*, 15 min, 20°C), and decant the supernatant. Do not worry if a little of the charcoal is decanted. Discard the pellet.
7. Remove any residual charcoal by filtering through a 0.45 μm nitrocellulose filter with a cellulose prefilter. You might have to change filters because of blocking. The filtrate should be clear.
8. Add an equal volume of concentrated HCl to the filtrate and stir gently. White crystals should form in a light yellow fluid.
9. Collect the precipitate on Whatman #1 and transfer to a baking dish previously cleaned with HCl. Chop up the precipitate into small pieces with a clean glass rod. All surfaces touching DABA must be very clean from this point.
10. Heat the open baking dish at 60°C in a fume hood overnight, in the dark. This may be done by covering a thermostatted waterbath on all sides with polyethylene sheet before adding the dish. This prevents evaporation of the water and protects the metal

parts of the waterbath from HCl vapor. The dish must be uncovered to allow HCl evaporation. The DABA is ready when there is no more HCl odor.

11. Store powdered DABA tightly sealed in a brown jar at −20°C. Allow the jar to warm to room temperature before opening.

## *Sample Preparation: Tube Method*

Two methods of sample preparation are given. The tube method is preferred if the sample is available, salt-free, in a volume of 100 μL or less. The sample may be crude and does not have to be in solution, e.g., a suspension of whole cells. If the sample is too dilute, too salty, or if you wish to remove soluble DABA-positive material, then you should precipitate the sample first by using the filter method described below (Note 1). The tube method gives lower background.

1. Spot the sample in the bottom of a 12 × 77 mm polypropylene tube and dry it at 50°C. The dried samples are stable for several days at room temperature, so you may accumulate several samples to assay later.
2. Prepare six DNA standards, including a zero DNA standard, in the same way as step 1, from the DNA solutions. Also prepare a blank sample with no DNA, but with the manipulations and buffers you use in your experimental samples (Note 2). The amounts of DNA in standards should bracket your experimental values.

## *Sample Preparation: Filter Method*

1. Spot the sample in a 12 × 77 mm polypropylene tube and add 0.2 or 0.5 mL of precipitation buffer. The final nucleic acid concentration should be at least 50 μg/mL, mostly yeast RNA from the precipitation buffer.
2. Mix, then add TCA to 5 or 10% (w/v) final concentration, mix again, and chill on ice for 30 min.

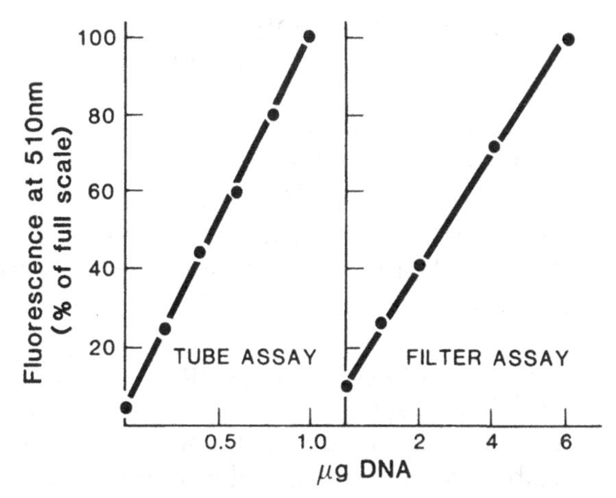


Fig. 1. Salmon sperm DNA was prepared and assayed by the tube method and the filter method.

3. Filter the sample onto a GF/C filter, rinsing the tube and apparatus three times with 2 mL of cold 1*N* HCl. Remove the chimney from the apparatus and rinse the filter once with about 1 mL of 95% ethanol.
4. Remove the filter, wet with ethanol, to a flat-bottomed glass vial and dry it there with the top off. The dry samples are stable at room temperature for several days.
5. Prepare DNA standards in the same way, on GF/C filters. Standards prepared by the tube method have a lower background than those prepared by the filter method (*See* Fig. 1).

## The DABA Assay

1. Take the powdered DABA out of the freezer and let it warm to room temperature. For each tube sample, you will need 32 mg of DABA powder plus 80 μL of $H_2O$. For each filter sample, you will need 80 mg of DABA powder plus 200 μL of $H_2O$ (Note 3).
2. Weigh the DABA, dissolve it in the appropriate volume of water, and add 0.1 mL of this solution to each tube or 0.25 mL to each filter.
3. Incubate the samples at 55–57°C for 45 min (Note 4), then dilute the samples with 1–3 mL of 1*N* HCl (Note 5).

4. Shake the filters gently to elute material from them. The samples are stable at room temperature for at least 1 d.
5. Turn on the fluorometer and let it warm up until the light source is stable (*see* Note 6).
6. Pour your highest DNA standard sample into the cuvet and adjust the sensitivity to give a full-scale reading (Note 7). Using the same instrument settings, measure fluorescence of all the other samples including the blank.
7. Plot a calibration curve, as in Fig. 1. This curve will establish the sensitivity of the assay. Save the samples until you have finished your data analysis; you may wish to read some samples again. In our hands, the assay is linear beyond 30 μg DNA, so that if you find an unexpectedly high experimental reading, you may extrapolate your standard curve on a scale of lowered instrument sensitivity. It is best to choose standards that span the data, however.
8. If we set the lower limit of sensitivity at 1.5× background, then the tube method is useful to 30 ng DNA and the filter method to 400 ng DNA. The sensitivity of the tube assay could be improved with purer DABA. The filter assay would be improved by using small filters, possibly very small nitrocellulose filters.

## Notes

1. In some cells, up to 50% of DABA-positive material is acid-soluble, and therefore, is probably not DNA (our own observations). Hence the need to precipitate the sample prior to sample preparation.
2. If there is salt present in your samples, then prepare the reference DNA samples in the same salt. Salt can quench the signal and add variability (*2*).
3. The DABA concentration in the assay mixture is about 1.6*M*.
4. Incubation during the assay is done most easily in uncovered tubes or vials in a covered waterbath. We have found that background rises at temperatures over 57°C. The reaction is incomplete below 50°C (*1*).

5. The volume of 1*N* HCl used to dilute samples following incubation is determined by the fluorometer. Use the minimum volume compatible with accurate readings.
6. You should get to know your fluorometer with a set of DNA standards before committing experimental samples. At this time you will also test the quality of your DABA and the sensitivity of the assay in your hands.
7. With filter samples, put only the HCl-eluate, not the filter, in the cuvet. The eluate will be slightly turbid.
8. In limited experiments, 25 mm nitrocellulose filters, 0.45 μm, have been substituted for GF/C filters with similar results except for a much slower flow rate during filtration.

## Acknowledgments

We thank Ellen Hughes for introducing us to the assay. This work was sponsored by USPHS Grants GM 26137 and CA 21797.

## References

1. Kissane, J. M., and Robins, E. (1958) The fluorometric measurement of deoxyribonucleic acid in animal tissues with special reference to the central nervous system. *J. Biol. Chem.* **233,** 184–188.
2. Hinegardner, R. T. (1971) An improved fluorometric assay for DNA. *Anal. Biochem.* **39,** 197–201.

# Chapter 3

# Preparation of "RNase-Free" DNase by Alkylation

*Theodore Gurney, Jr. and Elizabeth G. Gurney*

*Department of Biology, University of Utah, Salt Lake City, Utah*

## Introduction

Characterization of RNA molecules by electrophoresis or hybridization frequently requires nucleic acid concentrations over 1 mg/mL. High molecular weight DNA in a mixture of nucleic acids limits the solubility and interferes with electrophoresis. DNase treatment makes the mixture more soluble, even if DNA degradation is only partial.

The DNase, of course, must have no RNase activity. Most commercial purified pancreatic DNase I sold as "RNase-free" is not satisfactory (*1,2*). RNase activity is measured by production of acid-soluble ribonucleotides, but the assay is simply not sensitive enough for studies of high molecular weight RNA, because the few breaks that ruin large RNAs give no RNase activity. A more critical assay should be based on the integrity, after DNase treatment, of large RNA molecules.

A convenient test substrate for a more sensitive assay is the mixture of nucleic acids extracted from mammalian tissue culture cells labeled for 2 or 3 h with $^3$H-uridine, which labels both RNA and DNA. After DNase treatment, the mixture is analyzed by formaldehyde–agarose electrophoresis plus fluorography. High molecular weight RNA and DNA are well-separated in 0.7% agarose. A satisfactory test shows disappearance of the high molecular weight DNA and the simultaneous undiminished presence of the high molecular weight RNA species.

RNase A is a likely contaminant of pancreatic DNase I (*1*). Fortunately, RNase A can be inactivated by alkylation of a histidine in the active site (*3,4*). We have found alkylation satisfactory, using methods of Zimmerman and Sandeen (*1*), but others have not (*2*). Alternate approaches to this same problem use differential adsorption of RNase to a solid support (*2,5*). All methods should be tested using a sensitive assay method, such as the one described here, since different commercial preparations of DNase may have different RNase activities.

## Materials

1. DNase Sigma product number D5010, formerly DN-CL, was the most satisfactory of four commercial preparations tested. Sigma product number D4763 lost nearly all DNase activity during alkylation and was therefore not satisfactory. Worthington DPFF and PL Biochemicals 0512 both gave RNase-free DNase after treatment, but treated DNase was less stable than Sigma D5010. Enzymes were purchased in quantities of 10 mg lyophilized powder and were stored at −20°C.
2. 2.5 m*M* HCl: 3.2 L/preparation. Store at 4°C.
3. 1*M* sodium iodoacetate: 0.75 mL is prepared at the time of use.
4. DNase digestion buffer: 0.2*M* sodium acetate, pH 6.9; 10 m*M* $MgSO_4$. Autoclave and store at 4°C.

5. DNase assay substrate: 40 μg/mL purified undenatured DNA; 5 m*M* $MgSO_4$; 0.1*M* sodium acetate, pH 5.0. Store at −20°C.
6. Cell lysis buffer: 150 m*M* NaCl; 1 m*M* EDTA; 30 m*M* Tris-HCl, pH 7.3; 0.5% sodium dodecyl sulfate. Store at room temperature.
7. Phenol–chloroform: 50 mL distilled phenol (stored at −20°C in 50 mL aliquots, melted at 45°C when used), plus 50 mL chloroform, 1 mL isoamyl alcohol, 0.2 mL 2-mercaptoethanol, and 100 mL 1 m*M* EDTA, 10 m*M* Tris-HCl, pH 7.5. The liquids are mixed at room temperature and stored at 4°C. Two phases will separate; the lower one is phenol–chloroform. It is stable for at least 2 months at 4°C. Discard if it turns yellow. Seal the bottle with Parafilm (phenol–chloroform attacks plastic bottle tops). Work in a fume hood and avoid skin contact.
8. Self-digested pronase is prepared as follows (*6*): Commercial pronase, "grade B," is dissolved at 2 mg/mL in 0.5*M* $Na_4$ EDTA, pH 9.0, and incubated 2 h at 37°C. It is stored in quantities of 1 mL at −20°C and is not sensitive to freeze–thaw.
9. Pronase-SDS is prepared immediately before use by mixing one part of self-digested pronase and 19 parts cell lysis buffer. The final pH should be 8.0.
10. Ethanol is used as solutions of 95 and 70% (v/v) and is stored at 4°C.
11. Tissue-culture materials are:

    (a) Dulbecco's modified Eagle's medium
    (b) Calf serum
    (c) Physiological saline
    (d) 35 mm (8 $cm^2$) tissue culture-grade Petri plates
    (e) HeLa S-3 cells
    (f) 5,6-$^3$H uridine, 20–50 Ci/mmol.

# Methods

## *The Alkylation of DNase*

1. Alkylation of DNase is carried out as described by Zimmerman (*1*) with a few modifications. Ten milli-

grams of commercial DNase is dissolved in 2 mL of 2.5 m*M* HCl, rinsing out the bottle.

2. The solution is dialyzed for a few hours at 4°C against 1 L of 2.5 m*M* HCl, with stirring, then overnight against 1 L of fresh 2.5 m*M* HCl at 4°C with stirring.
3. Mix the following: 2.5 mL of 0.2*M* sodium acetate, pH 5.3, 2.0 mL of dialyzed enzyme, and 0.75 mL of 1 *M* sodium iodoacetate (freshly prepared), and incubate for 60 min at 55°C. A precipitate will form.
4. Dialyze this solution overnight against 1 L of 2.5 m*M* HCl at 4°C. Use sterilized pipets. Following the overnight dialysis, centrifuge for 30 min at 10,000*g*, 0°C. A swinging bucket rotor such as the Sorvall HB-4 is best. Use a sterilized glass tube. Carefully pipet off the supernatant into a sterilized tube with a tight cap. The protein concentration should be 3–4 mg/mL.
5. Store the sample at 4°C. The activity is stable for at least 2 yr, unfrozen.

## The Assay for DNase Activity

1. The assay for DNase activity is that of Kunitz (7). Dilute the DNase stock to 10% in water.
2. Determine $A_{280}$ and compute an approximate protein concentration, assuming $E_{280}^{1\%} = 11.1$.
3. Find two matched cuvets for a double-beam spectrophotometer. Into one, mix 1 part water and 5 parts DNase assay substrate. Into the other, mix 1 part of the 1/10 dilution (or some further dilution) of DNase stock and 5 parts DNase assay substrate at 25°C. Using the double-beam spectrophotometer, determine $\Delta A_{260}$ at 25°C as a function of time, taking readings every 30 or 60 s. The cuvet with DNase will show little change for 30–60 s; then its absorbance will rise to a maximum $\Delta A_{260}$ of about 0.16.
4. Determine the slope of the steep linear part of the curve, $\Delta A_{260}$ per min. One Kunitz unit of DNase activity causes a $\Delta A_{260}$ of 0.001/min/mL. A typical activity of a commercial DNase preparation is 2000 Kunitz units/mg protein, before and after alkylation.

## Preparation of Substrate for a Sensitive Nuclease Test

1. Seed HeLa cells at 5 × $10^5$ on one 35-mm Petri plate in Dulbecco's medium plus 10% calf serum. The cells should be at 1 × $10^6$/plate, rapidly dividing, at the time of labeling. You may wish to make duplicate cultures to determine cell numbers.
2. When the cells are at the right density, replace the medium with 1 mL of fresh medium (10% serum) containing 50 μCi $^3$H-uridine, and incubate for 3 h at 37°C.
3. Remove the medium and quickly rinse the plate twice in the cold room with 4°C physiological saline. In the cold room, drain the plate for a minute and remove residual saline.
4. Put on 0.4 mL of room-temperature Pronase-SDS. Warm the plate to room temperature and remove the viscous lysed cells to a 1500 μL microfuge tube, using a Pasteur pipet. Vortex vigorously to give a uniform suspension and then incubate the lysed cells at 40°C for 30 min.
5. Add 0.4 mL of the phenol–chloroform and vortex very vigorously to make a uniform emulsion. Break the emulsion by a brief (30 s) centrifugation and recover the upper aqueous phase which contains RNA and DNA.
6. Divide the aqueous material into about 10 samples of 40 μL in microfuge tubes. Add 100 μL of 95% ethanol to each tube, mix, and chill at least 2 h at −20°C. The substrate samples may be stored indefinitely at −20°C at this stage.
7. At the time of use, the samples will be centrifuged and washed. Each sample should contain about 5 μg DNA, 10 μg RNA, and $10^5$ cpm of incorporated radioactivity, precipated in ethanol.

## Test for RNase Activity and the Use of DNase

1. Centrifuge the substrate sample, precipitated in ethanol (8000*g*, 5 min, 4°C). Remove the supernatant.

2. Wash the precipitated material and the inside of the microfuge tube with 0.5 mL of 70% ethanol by vortexing at 4°C. Centrifuge again (8000*g*, 1 min, 4°C) and remove the supernatant. Drain the tube upside down in the cold for about 10 min and remove traces of ethanol from the inside walls with a fine-tipped pipet.
3. Dilute a working sample of DNase, from the stock solution, to 5 Kunitz units/mL in the DNase digestion buffer.

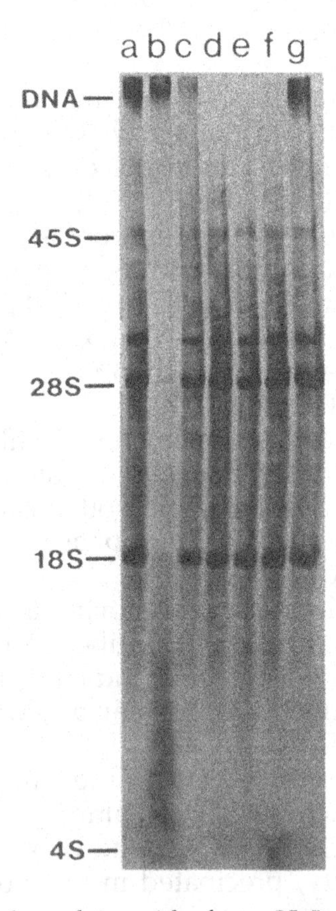


Fig. 1. $^{3}$H-labeled nucleic acids from HeLa cells after treatment with alkylated DNase and electrophoresis in 0.7% agarose, 2*M* formaldehyde: (a) untreated sample; (b) RNase-treated sample. Note that only DNA remains; (c) 0.04 U/mL DNase, as described in text; (d) 0.2 U/mL DNase; (e) 1.0 U/mL DNase; (f) 5.0 U/mL DNase; (g) untreated sample. The streak comes from a bubble in the sample slot.

4. To a test sample microfuge tube containing precipitated $^{3}$H-DNA and RNA, add 80 μL DNase digestion buffer and 20 μL diluted DNase. The final test concentrations are, in 0.1 mL (very approximately): 50 μg/mL DNA, 100 μg/mL RNA, and 1 Kunitz unit/mL DNase.
5. Incubate for 30 min at 4°C with occasional mixing.
6. Add 100 μL of Pronase-SDS, mix, and incubate at 40°C for 30 min.
7. Repeat the phenol–chloroform extraction and precipitation as in the preparation of substrate.
8. Dissolve the final precipitate in the triethanolamine sample buffer used with formaldehyde–agarose electrophoresis, described in Chapter 11.
9. Determine the radioactivity and analyze 1–2 × $10^4$ cpm by electrophoresis plus autoradiography.
10. Results of an assay are shown in Fig. 1, in which the DNase concentration was varied. Note that 0.2, 1, and 5 Kunitz units/mL were satisfactory. In other experiments, using 20 U/mL and 40 U/mL, we noted some RNA degradation, seen as selective reduction of 14 kb rRNA, compared to other RNA species. It is likely, therefore, that alkylation does not remove all RNase activity, but that satisfactory results can be obtained nevertheless by the proper choice of concentrations.

## Acknowledgments

We thank Jo-Ann Leong for introducing us to the alkylation procedure. This work was sponsored by USPHS Grants GM 26137 and CA 21797.

## References

1. Zimmerman, S. B., and Sandeen, G. (1966) The ribonuclease activity of crystallized pancreatic deoxyribonuclease. *Anal. Biochem.* **14,** 269–277.
2. Maxwell, I. H., Maxwell, F., and Hahn, W. E. (1977) Removal of RNase activity from DNase by affinity chromatography on agarose-coupled aminophenylphosphoryl-uridine 2′(3′)-phosphate. *Nucleic Acids Res.* **4,** 241–246.
3. Grundlach, H. G., Stein, W. H., and Moore, S. (1959) The nature of the amino acid residues involved in the

inactivation of ribonuclease by iodoacetate. *J. Biol. Chem.* **234,** 1754–1760.

4. Price, P. A., Moore, S., and Stein, W. H. (1969) Alkylation of a histidine residue at the active site of bovine pancreatic ribonuclease. *J. Biol. Chem.* **244,** 924–928.
5. Schaffner, W. (1982) Purification of DNase I from RNase by macaloid treatment, in *Molecular Cloning, a Laboratory Manual* (eds. Maniatis, T., Fritsch, E. F., and Sambrook, J.), p. 452. Cold Spring Harbor Laboratories Press, New York.
6. Kavenoff, R., and Zimm, B. H. (1973) Chromosome-sized DNA molecules from *Drosophila. Chromosoma* **41,** 1–27.
7. Kunitz, M. (1950) Crystalline deoxyribonuclease. I. Isolation and general properties, spectrophotometric method for the measurement of deoxyribonuclease activity. *J. Gen. Physiol.* **33,** 349–362.

# Chapter 4

# The Isolation of Satellite DNA by Density Gradient Centrifugation

***Craig A. Cooney and Harry R. Matthews***

*University of California, Department of Biological Chemistry, School of Medicine, Davis, CA USA*

## Introduction

The term satellite DNA is used for a DNA component that gives a sharp band in a density gradient and can be resolved from the broader main band of DNA in the gradient. The usual gradient material is CsCl in aqueous buffer and the $Cs^+$ ions form a density gradient in a centrifugal field. DNA in the solution sediments to its isopycnic point. The density of DNA is a function of base composition and sequence and so a homogeneous or highly repeated DNA sequence will form a sharp band in CsCl density gradients at a characteristic density. The resolution of this procedure may be enhanced or modified by binding ligands to the DNA. For example, netropsin binds specifically to A + T-rich regions of DNA and reduces their den-

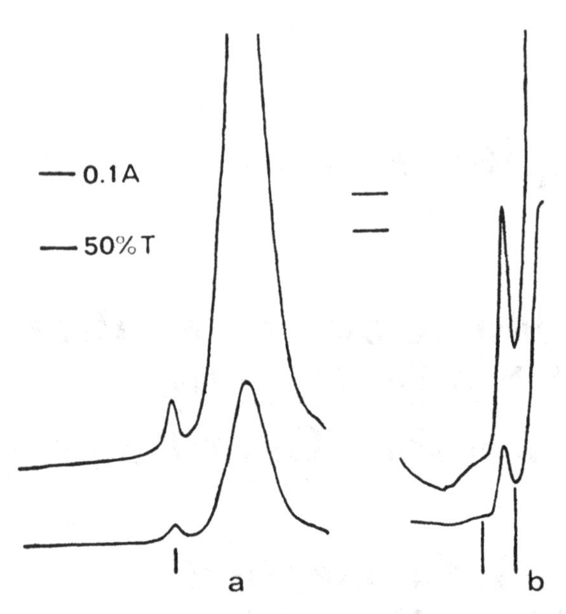


Fig. 1. Profiles from sections of 40 mL CsCl/bisbenzimide gradients. The upper profiles are $A_{276nm}$, where horizontal line marks are 0.1 *A* apart in a, b, and c and 0.25 *A* in d, relative to the lowest point on each absorbance profile. The lower profiles are $T_{276nm}$, where the lower horizontal line marks 50% T relative to the lowest point on each transmittance profile.
(a) Typical profiles of a gradient containing 500 μg of *Physarum polycephalum* $M_3C$ strain nuclear DNA, 400 μg (10 μg/mL) of bisbenzimide and 55% CsCl. The nuclear DNA contains about 1–2% of the G + C-rich extrachromosomal ribosomal RNA genes (rDNA). Note the especially good separation of rDNA (marked with a small vertical line) from the main band DNA when a relatively small amount of DNA is used.
(b) Typical profiles of a gradient containing 5 mg of *Physarum* $M_3C$ strain nuclear DNA, 2 mg (50 μg/mL) of bisbenzimide and 54% CsCl. The two vertical lines below the profiles border the region (fractions) of the gradient taken as rDNA, pooled with similar fractions of similar gradients (e.g., eight gradients total) and spun again.

sity (*1,2*). Another useful ligand is $Ag^+$, which must then be centrifuged in $Cs_2SO_4$ gradients to avoid precipitation of AgCl (*3*). Pharmacia has recently introduced $CsCF_3COO$ as a gradient material.

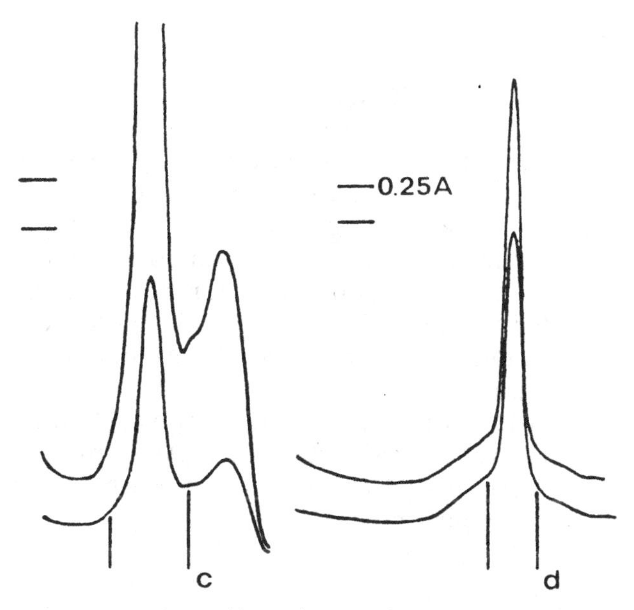


Fig. 1. (c) Typical profiles of a gradient containing fractions from gradients as in (b) spun in 54% CsCl with no additional bisbenzimide. The two vertical lines below the profiles border the region (fractions) of the gradient taken as rDNA, pooled with similar fractions of similar gradients (e.g., two gradients total) and spun again. (d) Typical profiles of rDNA fractions from gradients as in (c) spun in 55% CsCl with no additional bisbenzimide. The two vertical lines below the profiles border the region (fractions) of the gradient taken as rDNA, isopropanol-extracted, dialyzed, and ethanol-precipitated. Gradient centers in (a) and (d) (55% CsCl) are approximately at the position of the rDNA peak. Gradient centers in (b) and (c) (54% CsCl) are in the main band peak to the right of the rDNA peak. Gradient profiles, from left to right, are bottom to top, more dense to less dense.

In this chapter we describe the use of another DNA-binding dye with specificity for A-T-rich regions, bisbenzimide Hoechst 33258, to increase resolution in CsCl gradients or in KI gradients (*4,5,7*). We describe one good method for unloading the gradients after centrifugation; two other methods are described by Gould and Matthews (*6*). A typical fractionation, carried out in the presence of bisbenzimide is shown in Fig. 1.

## Materials

1. CsCl, 56% (w/w), in 10 m*M* EDTA, pH 7.5. A good reagent grade of CsCl is suitable for preparative work; check the CsCl concentration by measuring its refractive index, which should be 1.399.
2. An ultracentrifuge with a vertical rotor. An angle rotor may also be used, with somewhat extended run time, and a swing-out rotor may be used, but the latter is not usually recommended because of lower resolution and extended run times; note that CsCl attacks aluminum.
3. A tube unloading system. The system described here requires a peristaltic pump. An ultraviolet absorbance monitor and fraction collector are optional.
4. A refractometer.

## Method

1. Dissolve the DNA in 56% CsCl. If the DNA is in solution, add 1.28 g of solid CsCl per mL of solution. The volume will increase by 48% (*6*).
2. Check the refractive index of the solution. Add solid CsCl or 10 m*M* EDTA as necessary to adjust the density to give a refractive index of 1.399. The final density is 1.70 g/mL.
3. Load the centrifuge tubes as recommended by the manufacturer.
4. Centrifuge (30,000 rpm; 60 h; Beckman VTi 50 rotor; 20°C) to equilibrium.
5. Immediately unload the gradients, as follows:
   If a UV monitor is used, fill the flow path (tubing, flow cell, pump) of the fractionating equipment with stock 60% CsCl solution or dense liquid. Make sure the flow path is free of bubbles above and inside the monitor and zero the chart recorder.
6. Place the centrifuge tube in a rack and steadily lower a blunt-ended, hollow stainless-steel needle through the gradient until it reaches the bottom of the tube, then raise the needle about 1 mm from the bottom. Run the peristaltic pump backwards at its lowest

speed when inserting the needle into the gradient to prevent any bubbles from entering the flow path during needle placement.

7. Suck the gradient from the tube with a smooth peristaltic pump. If a multichannel pump is used, then multiple gradients may be unloaded simultaneously.
8. DNA may be recovered from the CsCl solution by dialysis and ethanol precipitation. If the DNA is to be rerun on CsCl, then simply add more CsCl solution, check the refractive index, and start the new run.

# Notes

1. If using bisbenzimide, then the CsCl solution should be 54% (w/w) CsCl, 10 m*M* EDTA, pH 7.5 with 10–50 μg/mL bisbenzimide and up to 0.1 mg/mL DNA giving a DNA:bisbenzimide ratio of about 2. Care must be taken to avoid the precipitation of DNA by bisbenzimide that occurs if concentrated solutions are mixed. We use bisbenzimide at 0.5 mg/mL in 10 m*M* EDTA, pH 7.5, and add it slowly to the DNA solution (about 0.1 mg/mL DNA in 54% CsCl) while swirling the DNA solution. Then, the CsCl concentration is adjusted to 54% (refractive index 1.395). Potassium iodide, 66% saturated, may be used instead of CsCl for the DNA + bisbenzimide procedure (*7*).
2. One of the crucial steps in this procedure is to get the initial CsCl density correct. If it is incorrect, then the DNA will band at the top or bottom of the tube. If the sedimentation pathlength is short, as in a vertical rotor, then the initial density is more critical than in the case of a longer pathlength, as in a swing-out rotor.
3. If using an angle rotor, it is usual to only partially fill the tube with the CsCl solution. It may then be necessary to fill the tube with glycerol to prevent it from collapsing during centrifugation. Aluminum tube caps will have a short life in a CsCl environment, so use stainless steel where possible. Do not get glycerol into the flow cell if you unload the gradient through a monitor.

4. The density gradient is formed by the centrifugal field. No significant time savings are achieved by preforming the gradient. Equilibrium is reached more quickly at high centrifugation speeds, but the bands are closer together because the density gradient is steeper. This is not a problem in an analytical ultracentrifuge because the bands are also sharper, giving good overall resolution. However, in preparative ultracentrifugation, some band broadening occurs during unloading of the gradients and so better final resolution is obtained at lower centrifugation speeds, limited by the time required to reach equilibrium. High loadings of DNA can also lengthen the run time required because of the high viscosity produced in the bands. High viscosity may also give problems during gradient unloading. See Note 1 for recommended DNA concentration.
5. Centrifuge at room temperature. DNA is stable in concentrated CsCl and gradient unloading is easier at room temperature. Cooling may lead to a potential precipitation problem at high CsCl concentrations and long pathlengths. Obey manufacturer's restrictions on rotor speeds with high density samples. Some protocols call for slow deceleration of CsCl gradients. This is not recommended since a large titanium rotor may take about 1 h to decelerate without braking. However, the rotor should not come to an abrupt stop and braking should cease at about 1000 rev/min, as is done automatically in some centrifuges.
6. If bisbenzimide and CsCl are used, then the DNA band(s) can be visualized by their fluorescence if the tube is illuminated with ultraviolet light. Remember to wear UV goggles.
7. The gradients are reasonably stable for up to 3 h at rest, especially if left in the rotor where they are protected from temperature fluctuations and unnecessary movement. However, they are generally less stable than sucrose gradients and must be handled carefully.
8. Gradients may also be collected very successfully by puncturing the bottom of the tube and either allowing the gradient to drip from the hole or displacing the gradient with 60% CsCl or another dense liquid. The equipment available commercially from ISCO Corp.

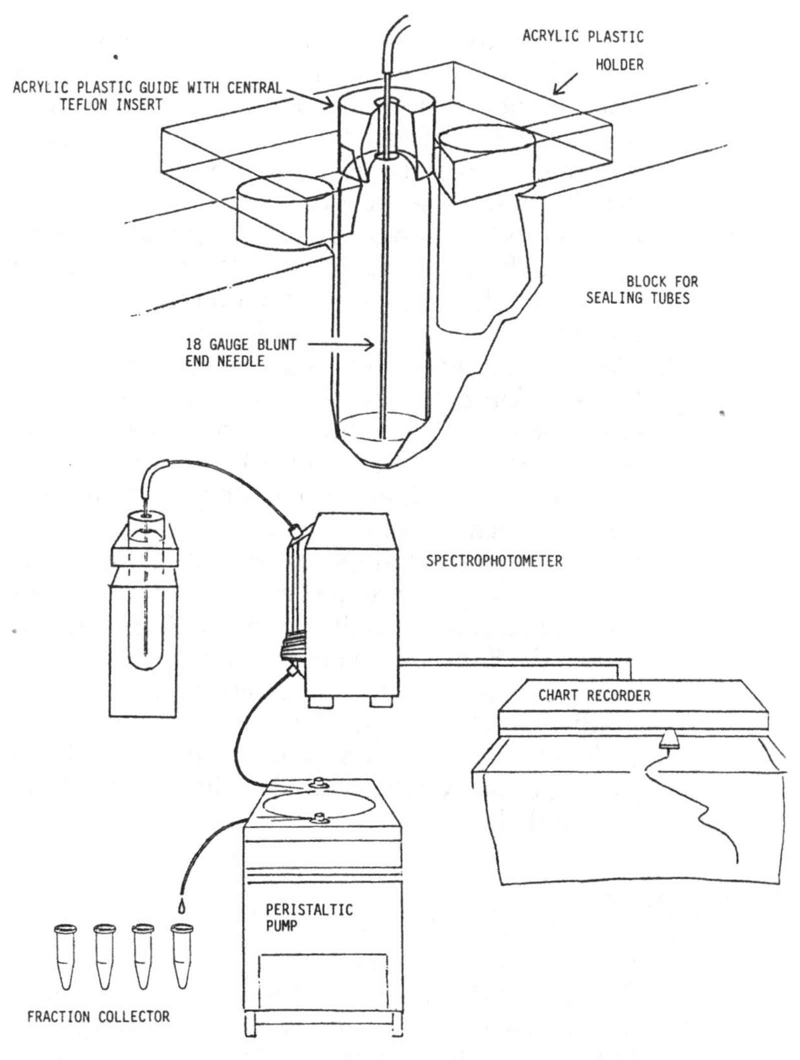


Fig. 2. Top. Cutaway diagram of equipment for holding the tube and needle. The tube is placed in a snug fitting rack. The tube sealing rack is used for Beckman Quick-Seal™ tubes. Quick-Seal™ tubes must have their sealed tops cut off to leave a small opening in the tube top. An acrylic plastic holder is placed over the tube and an acrylic plastic guide is placed in the holder, over the tube. The guide has a concave bottom to fit the rounded tube top and a teflon insert with a hole to snugly fit an 18 gage needle. Bottom. Equipment to monitor and fractionate gradients is arranged so that for most of the flow path the gradient contents are not inverted and the tubing inside diameter is small.

works well for the latter procedure. If a UV monitor is used, then CsCl gradients can be monitored at 254 nm, 260 nm, or longer wavelengths. However, KI is opaque at wavelengths below 270 nm and so a longer wavelength must be used, such as 276 nm or 280 nm. We currently use the procedure briefly described in the Method section using a "home-made" holder that fits over the centrifuge tube in the tube rack and provides a guide for lowering the hollow needle through the gradient. The set-up is shown in Fig. 2.

9. The guiding principles for the path of tubing carrying the solution from the centrifuge tube to the fraction collector are: (i) it should be short and not more than 1 mm internal diameter and (ii) the flow path should maintain the gradient orientation as far as possible, i.e., if collecting from the bottom of the tube, then the flow path should slope downwards. Particular care must be taken to prevent mixing caused by gradient reorientation in a monitor, if used. A flow rate on the order of 0.5 mL/min works well. If a monitor is used, it may be possible to fractionate only the part of the gradient that is of interest.
10. Bisbenzimide is removed from the DNA by isopropanol extraction of the DNA + CsCl solution before dialysis.

## References

1. Matthews, H. R., Johnson, E. M., Steer, W. M., Bradbury, E. M., and Allfrey, V. G. (1978) The use of netropsin with CsCl gradients for the analysis of DNA and its application to restriction nuclease fragments of ribosomal DNA from *Physarum polycephalum*. *Eur. J. Biochem.* **82,** 569–576.
2. Matthews, H. R., Pearson, M. D., and Maclean, N. (1980) Cat satellite DNA: isolation using netropsin with CsCl gradients. *Biochim. Biophys. Acta* **606,** 228–235.
3. Jensen, R. H., and Davidson, N. (1966) Spectrophotometric, potentiometric and density gradient ultracentrifugation studies of the binding of silver ion by DNA. *Biopolymers* **4,** 17–32.
4. Seebeck, T., Stalder, J., and Braun, R. (1979) Isolation of a minichromosome containing the ribosomal genes from *Physarum polycephalum*. *Biochemistry* **18,** 484–490.

5. Hudspeth, M. E. S., Shumard, D. S., Tatti, K. M., and Grossman, L. I. (1980) Rapid purification of yeast mitochondrial DNA in high yield. *Biochim. Biophys. Acta* **610,** 221–228.
6. Gould, H., and Matthews, H. R. (1976) *Separation Methods for Nucleic Acids and Oligonucleotides,* Elsevier/North Holland, Amsterdam.
7. Judelson, H. S., and Vogt, V. M. (1982) Accessibility of ribosomal genes to trimethyl psoralen in nuclei of *Physarum polycephalum. Mol. Cell. Biol.* **2,** 211–220.

# Chapter 5

# The Isolation of High Molecular Weight Eukaryotic DNA


## *C. G. P. Mathew*

*MRC Molecular and Cellular Cardiology Research Unit, University of Stellenbosch Medical School, Tygerberg, South Africa*


## Introduction

The isolation of high molecular weight eukaryotic DNA in good yield is an important prerequisite for the analysis of specific sequences by Southern blotting (Chapter 9), or for molecular cloning in phage or cosmid vectors (Chapter 49).

In the procedure described below, cells from the organism are disrupted by homogenization in Triton X-100 and the nuclei pelleted by centrifugation. The nuclei are then resuspended and treated with SDS, which dissociates the DNA–protein complex. The protein is removed by digestion with a proteolytic enzyme, proteinase K, and phenol extraction. Finally, the nucleic acids in the aqueous phase of the extract are treated with ribonuclease and the DNA is precipitated with ethanol.

Whole blood is a convenient source of DNA from larger mammals, but the procedure can easily be modified to isolate DNA from any cellular tissue.

## Materials

1. Cell lysis buffer: 320 m*M* sucrose; 1% (v/v) triton X-100; 5 m*M* $MgCl_2$; 10 m*M* Tris-HCl, pH 7.6.
2. Saline-EDTA: 25 m*M* EDTA (pH 8.0); 75 m*M* NaCl.
3. Sodium dodecyl sulfate: prepare a 10% (w/v) stock solution.
4. Proteinase K: prepare a 10 mg/mL stock solution.
5. 5*M* sodium perchlorate.
6. Phenol–chloroform: high-quality commercial phenol can be used without redistillation. Batches that are pink or yellow should be redistilled at 160°C to remove contaminants. Melt the phenol at 68°C, and add 8-hydroxyquinoline (antioxidant) to a final concentration of 0.1%. Mix with an equal volume of chloroform, and extract several times with 0.1*M* Tris-HCl, pH 8.0.
7. Chloroform: isoamyl alcohol (24:1).
8. TE buffer: 10 m*M* Tris-HCl (pH 7.5); 1 m*M* EDTA.
9. 20 × SSC: 3.0*M* NaCl; 0.3*M* sodium citrate, pH 7.0.
10. Ribonuclease: dissolve pancreatic ribonuclease A at 5 mg/mL in 10 m*M* Tris-HCl, pH 7.5. Heat at 80°C for 10 min.
11. Ethanol: 70% (v/v) and absolute.

Store solutions 1 and 6 at 4°C, and 4, 10, and 11 at −20°C. All other solutions may be stored at room temperature.

## Method

The procedure given below is for isolation of DNA from whole blood. It can be modified slightly to prepare DNA from cultured cells or tissues (*see* Note 1).

The method is based on that described by Kunkel et al. (*1*)

1. Collect 10 mL of whole blood in tubes containing EDTA or citrate as anticoagulant.

2. Add the blood to 60 mL of cell lysis buffer at 4°C and homogenize in a Potter homogenizer with a loose-fitting pestle. Pellet the nuclei by centrifugation at 2500*g* for 20 min at 4°C.
3. Suspend the pellet in 8 mL of saline–EDTA, and add 0.8 mL of 10% (w/v) SDS. Vortex briefly.
4. Add 50 μL of the proteinase K solution and incubate at 37°C for 2–4 h.
5. Add 0.5 mL of 5*M* sodium perchlorate and 8 mL of phenol–chloroform. Mix gently until homogeneous, and separate the phases by centrifugation at 12,000*g* for 10 min at 10°C.
6. Remove the aqueous phase with a wide-bore pipet, and extract it with an equal volume of chloroform–isoamyl alcohol. Mix gently, and separate the phases as in step 5.
7. Precipitate the DNA from the aqueous phase by adding 2 vol. of cold absolute ethanol (*see* Note 2). Lift out the precipitate with the sealed end of a Pasteur pipet, and shake into 1 mL of TE buffer. Dissolve overnight at 4°C, with gentle mixing.
8. Add 100 μL of 20 × SSC and 10 μL of ribonuclease, and incubate for 1 h at 37°C.
9. Add 2 mL of sterile distilled $H_2O$, and extract the solution twice with chloroform–isoamyl alcohol.
10. Precipitate the DNA by adding 2 vol. of absolute ethanol, and centrifuge at 5000*g* for 5 min at 5°C. Wash the pellet with 70% ethanol and dry under a vacuum. Dissolve the DNA in 0.5 mL of sterile distilled $H_2O$.
11. Scan a dilution of the DNA from 220 to 300 nm (*see* Note 3). Determine the absorbance at 260 nm and calculate the DNA concentration by assuming that the $A^{\%}_{1\,cm,260}$ is 200 [i.e., a 1 g/100 mL solution in a 1-cm light path has an absorbance of 200 at 260 nm (2)]. The yield of DNA from 10 mL of human blood is 200–400 μg.

## Notes

1. DNA can be isolated from cultured cells as follows: suspend cell pellet in tissue culture medium at a concentration of $10^7$ cells/mL. Add 5 vol of cell lysis buffer

and homogenize (Step 2). Continue with the procedure described for whole blood, but scale the volumes up or down according to the volume of cell lysis buffer used.

To isolate DNA from tissues, mince the tissue and blend it in liquid nitrogen, using a stainless-steel Waring blendor. Let the liquid nitrogen evaporate and add the powder to approximately 10 vol of lysis solution (3). After pelleting the nuclei (Step 2), follow steps 3–11 as described in the Method section.

2. In order to prepare DNA of very high mw ( > 30 kb), mixing with phenol and chloroform should be done very gently, and the DNA should not be ethanol-precipitated (3). Substitute precipitation steps 7 and 10 with extensive dialysis against several changes of TE buffer. In this case, the final product should not migrate faster than intact λ-DNA on a 0.4% agarose gel.
3. The scan of the DNA will detect impurities such as protein contamination, which can be removed by repeating the phenol–chloroform extraction, or traces of phenol, that will be removed by repeated extractions with chloroform–isoamyl alcohol.
4. Glassware and plasticware, such as Eppendorf tubes and automatic pipet tips, should be sterilized by autoclaving.

## References

1. Kunkel, L. M., Smith, K. D., Boyer, S. H., Borgaonkar, D. S., Wachtel, S. S., Miller, O. J., Breg, W. R., Jones, H. W., and Rory, J. M. (1977) Analysis of human *Y*-chromosome-specific reiterated DNA in chromosome variants. *Proc. Natl. Acad. Sci. USA* **74,** 1245–1249.
2. Old, J. M., and Higgs, D. R. (1983) Gene Analysis. In *Methods in Hematology. The Thalassaemias* (ed. Weatherall, D. J.), p. 78. Butler & Tanner, Rome and London.
3. Maniatis, T., Fritsch, E. F., and Sambrook, J. (1982) *Molecular Cloning: A laboratory manual.* Cold Spring Harbor Laboratory, New York.

# Chapter 6

# Preparation of Lyophilized Cells to Preserve Enzyme Activities and High Molecular Weight Nucleic Acids

*Theodore Gurney, Jr.*

*Department of Biology, University of Utah, Salt Lake City, Utah*

## Introduction

This procedure yields thin flakes of freeze-dried material from tissue culture cells. Up to 0.5 g of wet cells can be processed at one time. Dried cells, stored indefinitely at −20°C, have full lactate dehydrogenase activity (*1*) and DNA polymerase activity (*2*). The dried cells stored for a few days at room temperature also have apparently undegraded nucleic acids (*3*).

The procedure uses the first few steps of non-aqueous cell fractionation (*4*). Concentrated cells are frozen in melting Freon-12 and then dried while being refrigerated at −20 to −30°C.

## Materials

1. A high vacuum pump. The pumping speed should be at least 60 L/min and the ultimate vacuum should be lower than 50 mtorr.
2. A cold trap. A dry ice–methanol cold trap is placed in series between the pump and the sample. The trap requires a straight-sided Dewar flask to hold the dry ice and the methanol. *See* Fig. 1.
3. A vacuum gage. A thermocouple vacuum gage is used to determine the end point of the drying and to find vacuum leaks. The gage is placed between the cold trap and the sample.
4. Large test tubes (3 × 20 cm). The cells are frozen and dried in these tubes. The tubes must have a vacuum-tight seal to attach them to the vacuum system, and further, the seal must be tight at −20 or −30°C during drying. A chemistry glass shop can prepare 3 × 20 cm tubes with standard female Pyrex 29/42 tapered ground glass joints that attach to a male joint of the vacuum system. Silicone vacuum grease that does not

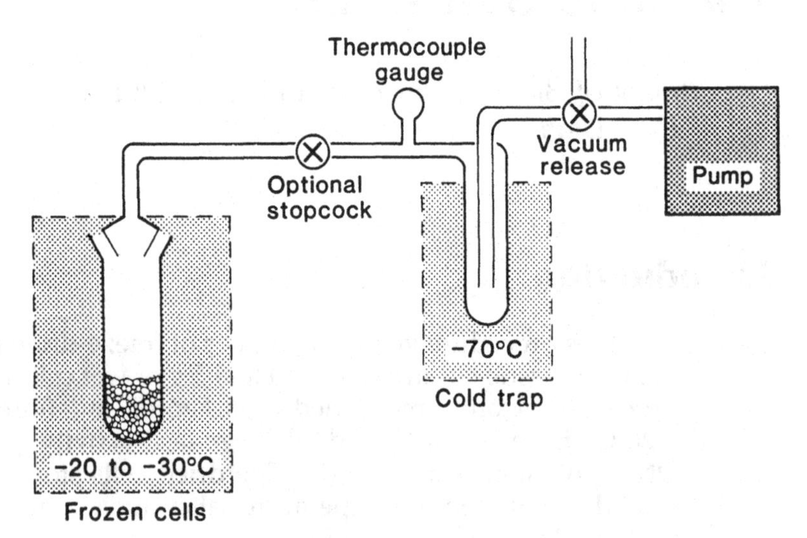


Fig. 1. Schematic diagram of the vacuum system. Not drawn to scale. The cold trap is chilled in a dry ice–methanol bath. The frozen cells are chilled in a refrigerated methanol bath (the bottom two-thirds of the tube only) or in a freezer.

freeze at −30°C should be used. Also, you will need rubber stoppers for the tubes.

5. Dewar flasks. Three straight-sided Dewars, one for the cold trap described above and two for freezing cells. The size of the Dewars should be cylindrical: 15 cm diameter and 20 cm in length, inside dimensions.
6. Cooling apparatus. The tubes of cells are chilled at −20 to −30°C during drying (Note 1). This can be done using a refrigerated methanol bath such as the "Multicool" (FTS Systems, Stone Ridge, NY). Lacking that, the chilling may be done in a freezer or the freezing compartment of a refrigerator. The door gasket of the freezer is cut to allow passage of the vacuum line. The piece of cut gasket is taped in place again between lyophilizations.
7. Dry box. Dried cells are handled in a cold $CO_2$ atmosphere made by putting dry ice in the bottom of a large styrofoam box, approximately 30 × 30 × 30 cm. Cold $CO_2$ will displace still room air with the lid of the box removed. Temperature can be measured by taping an alcohol thermometer to the inside of the box.
8. Heavy gloves and forceps. You will need to protect your hands against glass chilled to dry ice and liquid nitrogen temperatures.
9. Face mask. To be used when working with liquid nitrogen and frozen Freon-12.
10. Liquid nitrogen. About 500 mL per preparation.
11. Solid $CO_2$. "Dry Ice." About 1 kg per preparation.
12. Freon-12. About 100 mL per preparation. $CCl_2F_2$ is sold by refrigeration supply stores in cans of 400 g liquid, under pressure. You will need a dispensing valve and nozzle to fit the can. At one atmosphere, Freon-12 boils at −30°C and melts at −158°C. In the cans, Freon-12 can be stored at room temperature. In an open container, Freon-12 can be stored in a liquid nitrogen refrigerator. The Freon-12 may be recovered nearly completely after use and distilled by condensing it on dry ice at 1 atm.
13. Phosphate-buffered saline (PBS) (sterile): 140 m*M* NaCl; 2.7 m*M* KCl; 8.1 m*M* $Na_2HPO_4$; 1.5 m*M* $KH_2PO_4$; 0.9 m*M* $CaCl_2$; 0.5 m*M* $MgCl_2$. The ingredients are mixed, lacking calcium and magnesium salts,

and then autoclaved. After cooling, a sterile 100× calcium–magnesium salts mixture is added slowly while stirring. PBS is stored at +4°C.

14. Trypsin solution. This is used to suspend monolayer cells. The working concentration is 100 μg/mL of purified trypsin in PBS. The solution is stable for 2 months at 4°C. A concentrated stock solution, 10 mg/mL in 1 m*M* HCl, is stable indefinitely at −20°C.
15. Soybean trypsin inhibitor (SBTI). This is used to inactivate trypsin. The working concentration is 20 μg/mL in PBS. The solution is stable for 2 months at 4°C. A concentrated stock solution, 1 mg/mL in PBS, is stable indefinitely at −20°C.

# Methods

## *Preparation and Freezing of Cells*

1. To prepare 15 mL of frozen Freon-12, pour 200 mL of liquid nitrogen into a Dewar and chill a 3 × 20 cm tube in it. Using a nozzle or a rubber tube, introduce Freon-12 gas or liquid into the chilled tube. Introduce gas slowly so that it condenses. Store the tube with frozen Freon-12 on liquid nitrogen.
2. The preparation of concentrated cells should be carried out in a cold room at +4°C. Suspension cultures are chilled by pouring the warm culture onto one-half the culture volume of frozen PBS. The chilled cells are washed twice in cold PBS by centrifugation (2000*g*, 2 min, 4°C). The final pellet of cells is resuspended by adding one pellet volume of PBS or water (Note 2). The cells are chilled, washed, and frozen as rapidly as possible (*see* step 4 below)
3. Monolayer cultures are chilled by pouring off warm medium and quickly rinsing the monolayers twice with cold PBS. The cells are then made detachable, though not yet detached, by trypsin treatment in which washed monolayers are covered with trypsin solution for 2–10 min at 4°C (Note 3). HeLa cells require 4 min and 3T3 cells require 6–10 min, depending

on the growth rate; contact-inhibited cells need the longer treatment. The trypsinized, but still attached, cell sheets are then rinsed gently twice with SBTI and twice with plain PBS at 4°C. The cells are finally detached in PBS from the culture surfaces by vigorous pipetting and centrifuged once (2000*g*, 2 min, 4°C). The pellet of cells is suspended in a total of two pellet volumes of PBS or water and frozen immediately as described in the next step.

4. During the last 5 min of concentrating the cells, transfer the tube of frozen Freon-12 from liquid nitrogen to a room temperature or 4°C Dewar, in order to begin thawing it. The Freon should take 3–5 min to become half-melted. In the cold room, draw the concentrated cells into a Pasteur pipet and drip the cell suspension into the melting Freon-12. The cells must drop directly into the Freon-12 and not hit the walls of the tube. At the end of the dripping, there should be a little solid Freon-12 remaining unmelted (Note 4). Put the tube of frozen cells plus Freon-12 in a Dewar with dry ice. Cover the tube with a rubber stopper and cover the Dewar with aluminum foil. The cells may be stored indefinitely at −70°C or colder.
5. Freon-12 may be recovered using the following method. Chill a 10 mL pipet with dry ice and use the chilled pipet to remove Freon-12 from the tube containing the frozen cells into a clean chilled (dry ice) 3 × 20 cm tube. You must remove nearly all of the Freon-12 or your cells will be blown out of the tube when you apply vacuum. The Freon-12 may be stored below its boiling point and used again. Cover the tube of frozen cells with a stopper and store the tube on dry ice.

## Lyophilization of Cells

1. The vacuum system should be set up and tested with an empty tube in place of the tube containing frozen cells. Record the vacuum after an hour of pumping; it should be below 50 mtorr. A good two-stage mechanical pump should go below 10 mtorr. This is the time to fix the leaks in the system.

2. When the vacuum system is ready, attach the tube containing frozen cells. Start with cells at −70°C and work quickly so as not to warm them above −20°C. Apply the initial vacuum slowly over a few seconds to avoid explosive boiling of the residual Freon-12.
3. Pump for several hours, or overnight, until the vacuum stops dropping and approaches the value you recorded using the empty tube, then release the vacuum, cap the tube, and store it at −20°C. Store the tube upright to avoid dried cells adhering to vacuum grease on the ground glass joint.
4. To transfer the dry cells, chill the tube further on dry ice to make the vacuum grease hard. In the dry box, pour the dried cells into your experimental container through a glassine paper funnel. If you wish to store dried cells for later Southern blot analysis, you may pour them into Nalgene sealable plastic bags. Squeeze the air out before sealing. The bags may be mailed at ambient temperature. The structure of dried cells is

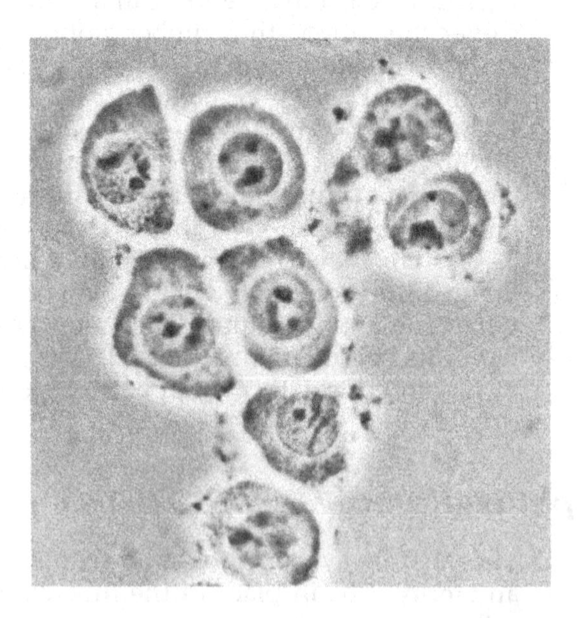

Fig. 2. Photomicrograph of HeLa S-3 cells grown in monolayer and lyophilized. The dried cells were suspended in hexylene glycol to make the photograph. The diameter of the nucleus is approximately 8 μm in the photograph.

apparently preserved, at the level of light microscopy (4) (see Fig. 2).

## Notes

1. The upper temperature limit (−20°C) for chilling cells during drying is determined by the stability of the activity you are trying to preserve. If your enzyme is more stable, you may use a higher temperature. Nucleic acids are stable at −20°C during lyophilization. I have not explored higher temperatures. The lower temperature limit (−30°C) is determined by the vapor pressure of water. Colder temperatures take too long to dry the cells.
2. Water is used to resuspend cells if more salts from PBS would affect later biochemistry. Enough salts are contributed by the pellet to keep the cells from lysing when suspended in up to three pellet volumes of water just before freezing.
3. The trypsin treatment of monolayer cultures has had no detectable effect on internal cell proteins and enzyme activities with this rinsing scheme. External proteins are vulnerable, however. The alternative approach of scraping cells off dishes often lyses cells.
4. Freezing cells in liquid Freon chills them faster than dipping them into liquid nitrogen. The temperature of melting Freon-12 is buffered at its melting point, −158°C. Liquid nitrogen is colder, but the freezing cells are insulated from it by $N_2$ gas because the nitrogen boils. Freon-12 might solubilize very hydrophobic molecules, but most biological molecules are certainly insoluble in it.

## Acknowledgment

This work was supported by USPHS Grant GM 26137.

## References

1. Gurney, T., Jr. and Collard, M. W. (1984) Nonaqueous fractionation of HeLa cells in glycols. *Anal. Biochem.* in press.

2. Foster, D. N., and Gurney, T., Jr. (1976) Nuclear location of mammalian DNA polymerase activities. *J. Biol. Chem.* **251,** 7893–7898.
3. Unpublished results: $^3$H-uridine labeling showed that 14 kb rRNA was undegraded and that labeled DNA was larger than 30 kb in single-stranded molecular weight.
4. Gurney, T., Jr. and Foster, D. N. (1977) Nonaqueous isolation of nuclei from cultured cells. *Methods in Cell Biology* **16,** 45–68.

# Chapter 7

# Agarose Gel Electrophoresis of DNA

**_Stephen A. Boffey_**

*Division of Biological and Environmental Sciences, The Hatfield Polytechnic, Hatfield, Hertfordshire, England*

## Introduction

This book contains many chapters describing methods for isolating and modifying DNA molecules. The most usual way of checking the success of such procedures is by looking at the products using electrophoresis in agarose gels. This process separates DNA molecules by size, and the molecules are made visible using the fluorescent dye ethidium bromide. In this way DNA can be checked for size, intactness, homogeneity, and purity. The method is rapid and simple, yet capable of high resolution, and is so sensitive that usually little of the sample is needed for analysis.

Agarose forms gels by hydrogen bonding when in cool aqueous solution, and the gel pore size depends on agarose concentration. When DNA molecules are moved through such a gel by a steady electric force, their speed of movement depends almost entirely on their size, the smallest molecules having the highest mobilities. Very large molecules are virtually immobile in high concentration gels, while small fragments will all move at the same rate in dilute gels; thus the gel concentration must be chosen to suit the size range of the molecules to be separated. Gels containing 0.3% agarose will separate linear double-stranded DNA molecules between 5 and 60 kilobases (kb) in size, whereas 2% gels are most satisfactory between 0.1 and 3 kb. We routinely use 0.8% gels to cover the range 0.5–10 kb. For calibration, when determining the sizes of DNA fragments, a straight line graph can be obtained by plotting mobilities against log molecular weights of suitable markers, although this linear relationship only holds over a limited range of DNA sizes for each gel concentration. Such calibrations cannot be extended from linear DNA to circular forms, since linear, open circle, and supercoiled forms of the same DNA will have markedly different mobilities.

Electrophoresis of DNA can be done in vertical or horizontal apparatus, in rods or slabs, using wicks, agar bridges, or (as described here) direct contact between gel and buffer. This chapter gives details of a horizontal slab system, similar to that described by Maniatis et al. (*1*), in which the whole gel is submerged in buffer during electrophoresis; it is often referred to as a 'submarine' or 'submerged' gel system. Owing to their ease of use, ability to support weak, dilute gels, and excellent performance, submerged gels are used widely for the electrophoresis of DNA. To avoid prolonged exposure to UV radiation it is usual to photograph gels for subsequent analysis, and so photography is also covered in this chapter.

## Materials

1. Electrophoresis apparatus. This can be bought ready made, but it is easy to make, and can be tailored to your own needs. It consists of three parts:

   (a) Casting plate. A glass plate 23 × 12.7 cm, 3 mm thick, with any sharp edges removed. Perspex plates tend to warp in use. A roll of zinc oxide tape, 12.5 mm wide, is needed to form a wall round the plate.

(b) Well forming comb. Cut from perspex 3 mm thick, giving 10 teeth, each 9 mm across, separated by 2.8 mm gaps. This is glued to supports at each end so that, when placed across the casting plate, the teeth are about 1 mm above the plate. **N.B:** The teeth must not touch the glass, or bottomless wells will result.

(c) Electrophoresis tank. Also made from perspex. This should be just wide enough to take the casting plate plus two layers of zinc oxide tape. To avoid any danger of the running buffer becoming exhausted during electrophoresis, the tank is designed to hold a large volume of buffer, but has a relatively shallow central section where the gel sits. Figure 1 shows a typical tank. The dimensions can, of course, be altered to suit particular needs. It is advisable to incorporate a microswitch that will cut off the power supply if the lid is removed.

2. A power supply that can produce direct currents up to about 100 mA, and constant voltages up to 100 V will be needed. (The gel described here is normally run at 40 V, drawing about 40 mA.) The output should be of the 'floating' type (i.e., not grounded), and must be protected against short-circuits by a fuse or other overload protection.
3. Transilluminator. It is worth buying the most powerful UV transilluminator you can afford if you want to obtain the highest possible sensitivity. An emission peak near 300 nm gives high sensitivity, yet minimizes damage to DNA by photonicking. You will need safety goggles (ordinary spectacles are not adequate) to protect your eyes, and a 5 mm thick perspex screen to protect your face from being sunburnt.
4. Camera. The Polaroid MP-4 camera cannot be beaten for quality and flexibility. It should be fitted with an orange/red Wratten 23A filter, above a perspex or glass filter to block UV light (the Wratten filter fluoresces in UV light). Type 667 film is highly sensitive and gives good quality prints; type 665 is slower, but produces

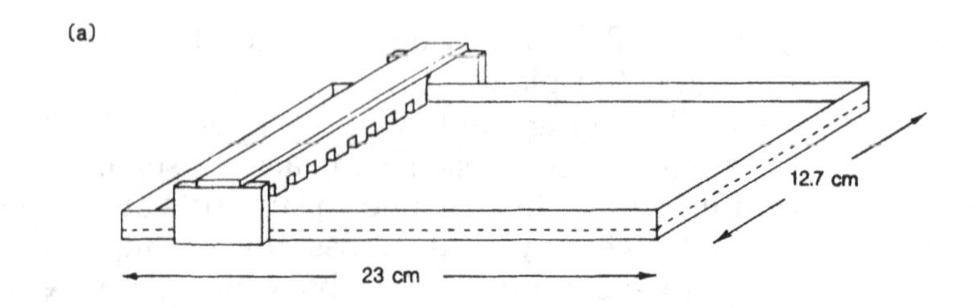


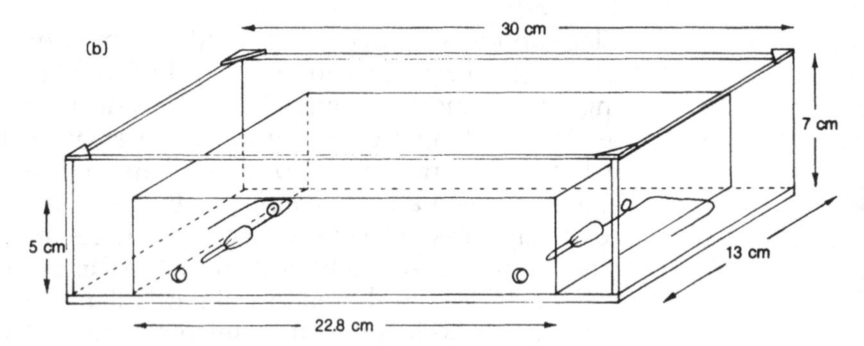


Fig. 1. (a) Glass casting plate with zinc oxide tape forming a wall around it. The well-forming comb, attached to supports, is in place near one end of the casting plate; note that the teeth of the comb do not quite touch the glass plate. (b) Electrophoresis tank, with raised central platform to support gel. Electrodes are made of platinum wire. Even though the voltages used tend to be low, this apparatus should always be run with a lid in place; ideally a microswitch should be fitted to cut off the power when the lid is raised.

both negatives and prints. Good results can be obtained using 35 mm (or larger format) cameras with a film such as Ilford FP4 and the same filters as above, but results are not available immediately.

5. Gel running buffer (stock solution): Tris 0.9*M*, $Na_2EDTA$ 25 m*M*, boric acid 0.9*M*, the whole being adjusted to pH 8.2 using HCl. This stock solution is ten times its final working concentration, and can be stored indefinitely at 4°C.
6. Agarose. Use an agarose with a low coefficient of electroendosmosis ($-m_r$), such as Type I (Sigma) or Seakem HGT (FMC Corporation, Rockland, USA). Add single-strength running buffer to weighed agarose powder to give the desired concentration (see Introduc-

tion), then heat in an autoclave, boiling water bath, or microwave oven to dissolve the agarose. Swirl the solution to ensure the agarose is uniformly distributed, and keep it at about 50°C until needed.

7. Gel loading solution. Ficoll (type 400) 30%, bromophenol blue 0.25% (both w/v) in single-strength gel running buffer. Include 0.5*M* EDTA if it is wished to use this as a 'stopping mix.' The solution can be stored indefinitely at room temperature.
8. Ethidium bromide, 5 mg/mL in single-strength gel running buffer. Wear disposable gloves when handling this powerful mutagen.

## Method

1. Wash the glass casting plate thoroughly. Alcohol can be used to remove any grease. Make sure the plate is completely dry, then form a wall round it using zinc oxide tape. Press the tape firmly against the edge of the glass to ensure firm attachment, paying particular attention to the corners of the plate and the region where the tape ends overlap. Do not stretch the tape, or it will bow in along the sides of the plate. This should result in a leakproof wall about 9 mm high all round the plate (see Fig. 1).
2. Place the prepared plate on a bench or leveling table, and check with a spirit level that it is perfectly horizontal.
3. Add 100 mL of buffer to the appropriate weight of agarose, and heat to dissolve the agarose. Allow the solution to cool to 50°C before pouring it all onto the casting plate, giving a thickness of about 3 mm in the apparatus described above. If the agarose is too hot, it may weaken the tape adhesive and leak from the plate; if it is too cool, it may gel unevenly on the plate. Agarose can be melted and gelled several times without any ill effects. As soon as the agarose is poured, place the well-forming comb in position about 3 cm from one end of the plate, ensuring that the comb is exactly at right angles to the sides of the plate and is free of air bubbles. Leave at room temperature until the gel has set; this

takes at least half an hour, and the gel looks cloudy when set.

4. Gently remove the comb and place the gel, still on its glass plate, in the electrophoresis tank, with its wells near the cathode. Now pour running buffer into the tank until its level is about 1 mm above the zinc oxide tape. Note that the tape has been left in place to prevent any movement of the gel off the glass plate during handling or electrophoresis; it has no effect on the electrophoresis.
5. Because samples must be loaded into the wells through running buffer, their densities must be increased to ensure that they fall into, and remain in the wells. Therefore, to each sample is added 0.1 times its volume of 'loading buffer.' Owing to the high density of loading buffer, care is needed to ensure that it mixes completely with the sample. If 0.5*M* EDTA is included, this solution can double as a 'stopping mix' to arrest restriction endonuclease digestions.
6. Samples are loaded into the wells using a micropipet or microsyringe. With the syringe or pipet tip a couple of millimeters above the well, gently dispense the sample, which will fall into the well. This method needs a reasonably steady pair of hands, but avoids any danger of accidentally injecting the sample into the gel beneath a well.
7. When all samples are loaded, the apparatus is closed, connected to a power pack, and run at 40 V overnight. After about 16 h, double-stranded DNA about 800 bp in length will have moved roughly 12 cm along a 0.8% gel.
8. Turn off and disconnect the power supply, and transfer the gel on its glass plate into a shallow tray. Remove the zinc oxide tape, and then cover the gel with 250 mL of gel running buffer containing 50 μL of 5 mg/mL ethidium bromide (wear gloves when handling this solution or stained gels). After about half an hour, the gel can be transferred onto a UV transilluminator for viewing; the glass casting plate must not be left in place at this stage since it will block the UV light. **N.B.**: Protect eyes and skin from the UV radiation. Nucleic acids on the gel will appear orange, owing to the fluorescence of bound ethidium bromide.

## Notes

1. It is usual to photograph a gel and to use the photograph for measurements of mobilities or interpretation of restriction patterns. This minimizes damage to nucleic acids by photonicking (an important consideration if they are to be recovered from the gel for further use), prolongs the life of the transilluminator filter, reduces the risk of sunburn, and may reveal bands that were too weak to be visible to the unaided eye.
2. If no bands are visible, incorrect polarity of electrodes might be to blame (does your power supply have a 'reverse polarity' switch?): always check to be certain that, after a few minutes, the bromophenol blue has started to move towards the anode. Perhaps the gel is poorly stained: check ethidium bromide concentration and allow a full half-hour for staining. Was enough DNA loaded? Less than 5 ng can be detected in a single band, but the more complex a sample (i.e., the greater the number of bands it produces on electrophoresis), the more of it will be needed to give visible bands.
3. Excessive background fluorescence is usually caused by unbound ethidium bromide. Transfer the gel to buffer or distilled water and leave for 30 min to wash out excess dye. Do not prolong this, or weak bands may disappear.
4. Streaking of bands along tracks is most commonly attributable to overloading, but can also be seen if the DNA has not completely dissolved before loading.
5. If bands are poorly resolved, and this is not a result of overloading, it may be possible to improve resolution by increasing the running time and/or changing to a more suitable agarose concentration (*see* Introduction). It is unwise to increase voltage gradients above 5 V/cm if high resolution is needed.
6. There are many ways in which this method can be altered to suit specific needs. Gels can be larger or smaller than described, and when reduced to less than about 5 × 10 cm, with proportionately smaller wells, can be used for rapid screening of samples. Such 'minigels' are usually run at 10–15 V/cm for 0.5–1 h.

   Agarose gels can be prepared in vertical electrophoresis tanks, as often used for polyacrylamide gel electrophoresis. These gels can give slightly sharper bands than horizontal types, and are preferred by some workers. However, they are unsuitable for low agarose concentrations, and lack the simplicity and reliability of horizontal systems.

Electrophoresis may be carried out in the presence of ethidium bromide. This eliminates a separate staining step, and if the casting plate and tank base are made of UV-transparent material, the movement of bands may be monitored during electrophoresis. However, intercalation of dye alters the running properties of DNA, and may even alter the order of linear and supercoiled bands along the gel.

7. This type of electrophoresis is essentially analytical, although it can be used to isolate microgram amounts of a particular DNA (*see* Chapter 10 for methods of recovering DNA from gels). For a detailed description of gel apparatus which can be used for the preparative fractionation of up to 50 mg of DNA; *see* Sealey and Southern (2).

## References

1. Maniatis, T., Fritsch, E. F., and Sambrook, J. (1982) *Molecular Cloning*: A laboratory manual. Cold Spring Harbor Laboratory, New York.
2. Sealey, P. G., and Southern, E. M. (1982) Electrophoresis of DNA, in *Gel Electrophoresis of Nucleic Acids: A Practical Approach*, (Eds. Rickwood, D., and Hames, B. D.), pp. 39–76. IRL Press, Oxford and Washington.

# Chapter 8

# Autoradiography of Gels Containing $^{32}P$

***Verena D. Huebner and Harry R. Matthews***

*University of California, Department of Biological Chemistry, School of Medicine, Davis, California*

## Introduction

Autoradiography of gels containing a $^{32}P$ label is used to detect and quantitate the radioactive label in a particular band or spot. Since $^{32}P$ is a high energy β-emitting isotope, no fluor is required in the gel (as in fluorography) to increase the efficiency of detection. However, since most of the radiation of high energy emitters (e.g., $^{32}P$ or $^{125}I$) passes through the film without being absorbed, it is common to employ an intensifying screen (e.g., calcium tungstate), where the radiation from the isotope is absorbed by a fluorescent compound that reemits the energy as light. The intensifying screen is placed against the film on the opposite side from the radioactive source. Any radiation that passes right through the film is then absorbed by the screen, where light is emitted back onto the film. In this way, much more of the emitted radiation is used and the film is exposed by a combination of direct and indirect autoradiography. The time needed to obtain an autoradiography is therefore reduced when an intensifying screen is used. If a small amount of label is present, it is also advisable to use an intensifying screen in the cassette to decrease the time of exposure required.

## Materials

1. Darkroom
2. Film cassette, plus intensifying screen, for higher sensitivity
3. Film, Kodak XS5
4. Automatic developer or baths with photographic developer, stop and fix solutions
5. Dry gel containing $^{32}P$-labeled samples.

## Method

1. Run, fix, stain if required, and dry the gel (*see* Vol. 1, Chapter 16).
2. In complete darkness, place the gel in an X-ray cassette. Lay one sheet of film on the gel. The intensifying screen, if used, can then be placed on the film. Close the cassette.
3. Place the cassette where it will not be exposed to other penetrating radiation, such as $^{125}I$ or other γ-emitting sources, X-ray sources, high amounts of $^{32}P$, or other high energy β-emitters. Unlike fluorography, the exposure may be carried out at room temperature, although if only a small amount of radioactivity is present, exposure at −70°C can increase the sensitivity approximately twofold.
4. In a darkroom, remove the film from the cassette and develop in an automatic X-ray developer.

## Notes

1. Any type of gel system is suitable since quenching is not a problem, unlike fluorography. The gel may be

stained, again because color quenching is not a problem. Thin gels may be autoradiographed wet, without serious loss of resolution or sensitivity. Note, however, that if this procedure is used for $^{35}S$ or $^{14}C$ autoradiography, then the gel must be dried. Generally, we dry the gel because it is then easier to handle.

2. A safety light may be used sparingly in the darkroom. It is important that the intensifying screen be placed next to the film for maximum effect. Radiation that passes through the film will interact with the screen, causing it to fluoresce, and this will expose the adjacent film. The use of an intensifying screen can decrease exposure time approximately twofold.
3. The cassette should be light-tight, so that further protection from light is not usually required. Fogging at the edges of the film usually indicates light leakage. Exposure time is about 4 h for 350–1000 dpm/cm$^2$ in the presence of an intensifying screen. Longer times may be used for lower amounts of radioactivity without background problems, but the short half-life of $^{32}P$ (14.3 d) limits the time available. For this reason, we recommend starting an autoradiography as soon as possible and using the longest exposure time that is likely to be needed. If necessary, an additional, shorter, exposure can then be carried out.
4. The film may also be developed by hand by placing it in a film holder and immersing it in D19 developer for 7 min, followed by the stop-bath for 0.5 min, the fix-bath for 3 min, and extensive rinsing in distilled water.
5. It is important to mark the film in at least three places so that it can be properly aligned with the gel. This is most easily done by using radioactive ink to mark the gel in three corners just prior to exposure. Remember that the two sides of the film may be identical so at least three well-spaced marks are essential.

# Chapter 9

# Detection of Specific DNA Sequences—The Southern Transfer

*C. G. P. Mathew*

*MRC Molecular and Cellular Cardiology Research Unit, University of Stellenbosch Medical School, Tygerberg, South Africa*

## Introduction

The purpose of this technique is the detection and characterization of specific DNA sequences. The DNA is fragmented by digestion with a restriction endonuclease, and the fragments separated by agarose gel electrophoresis. The DNA is then denatured in the gel and transferred to a nitrocellulose filter. This is incubated with a $^{32}$P-labeled probe, which is DNA having a base sequence complementary to the DNA that is to be detected on the filter. After hybridization of the probe to its complementary sequence, unbound probe is washed off. The position of the probe on the filter is then detected by autoradiography. This procedure was developed by E. M. Southern of Edinburgh University (*1*), and is generally referred to as the Southern transfer or Southern blot.

The technique has two types of application. Firstly, it is used routinely to screen and map recombinant plasmids or phages that have been generated by the cloning procedures described elsewhere in this book. Secondly, it is used to detect and characterize specific sequences in preparations of genomic DNA. The latter application requires the detection of picogram amounts of a particular DNA sequence among thousands or millions of other DNA sequences. The technique is therefore very sensitive and very specific. Theoretical aspects and applications of the techniques have been recently reviewed (2).

## Materials

1. The apparatus used for agarose gel electrophoresis has been described in Chapter 7. Either horizontal or vertical gels can be used. The horizontal gel is easier to pour, but the vertical gel is smooth and flat, which allows better contact with the filter.
2. The hybridization is done in a perspex chamber or hybridization box designed by Alec Jeffreys of Leceister University (*see* Note 1). The features and dimensions of the box are illustrated in Fig. 1. It can easily be constructed by University or Hospital workshops.
3. X-ray film and cassettes are required for the autoradiography. Films such as Kodak X-Omat R, Kodak XAR 5, or Fuji RX are suitable. The cassette should be fitted with a calcium tungstate intensifying screen such as Dupont Cronex lightning plus or Fuji Mach 2.
4. Restriction endonuclease buffers: These are prepared as a 10× stock, according to the manufacturer's instructions. Use sterile distilled water (sdw) and filter through a 0.45 μm cellulose acetate filter (e.g., Millex HA) (*see also* Chapter 31).
5. Nuclease-free bovine serum albumin (BSA), 5 mg/mL in sterile distilled water.
6. Electrophoresis buffer: Prepare a 10× stock solution containing 0.89*M* Tris-borate, 0.89*M* boric acid, and 0.02*M* EDTA.
7. Loading buffer: 0.1% (w/v) Orange G, 20% Ficoll, 10 m*M* EDTA (pH 7.0).

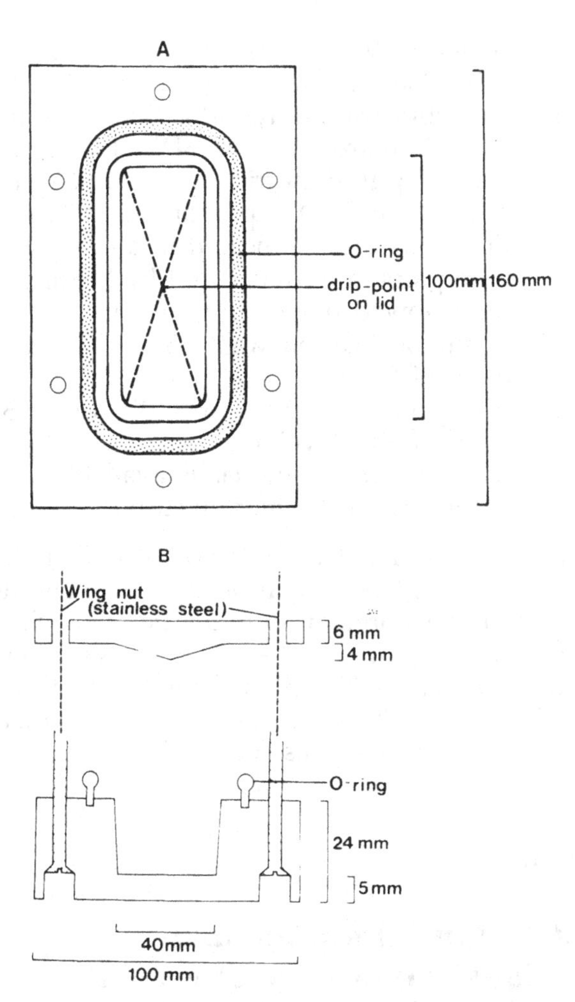


Fig. 1. Diagram of the hybridization chamber designed by Alec Jeffreys, showing the plane (A) and elevation (B).

8. Ethidium bromide: 10 mg/mL (w/v) in sterile distilled water (**N.B.**: ethidium bromide is mutagenic).
9. 0.25*M* HCl (optional)
10. 20 × SSC:, 3*M* NaCl, 0.3*M* Tri-sodium citrate, pH 7.6.
11. Denaturing solution: 0.5*M* NaOH, 1.5*M* NaC1.
12. Neutralization solution: 0.5*M* Tris-HCl in 20 × SSC, pH 5.5.
13. 3 × SSC: Dilute 20× stock.
14. 3 × SSC, 10× Denhardts. Prepare 100× Denhardt's stock containing: 2% (w/v) bovine serum albumin

(Fraction V, Sigma), 2% (w/v) polyvinylpyrrolidone (PVP-360, Sigma), 2% (w/v) Ficoll 400 (Pharmacia).

15. Prehybridization/hybridization solution: 3 × SSC, 10× Denhardt's, 0.1% SDS, 10 μg/mL polyadenylic acid, 50 μg/mL herring sperm DNA (*see* Note 3). Herring sperm DNA is prepared as a 2 mg/mL stock solution. It is then sonicated to an average length of 600 base pairs, and denatured by heating at 100°C for 10 m, followed by cooling on ice.
16. Posthybridization wash: 3 × SSC, 10× Denhardt's, 0.1% SDS.
17. Stringent wash: 0.1% SDS, 0.1–1.0× SSC (*see* Note 4).
18. X-ray film developer and fixer for the autoradiography are commercially available, and are made up according to the manufacturer's instructions.

Stock solutions of 1*M* Tris-HC1 (pH 7.5), 0.1*M* EDTA (pH 7.0), 1*M* $MgC1_2$, and 10% (w/v) SDS should be prepared and can be stored at room temperature. Stock solutions of restriction enzyme buffers, bovine serum albumin, polyadenylic acid, 100 × Denhardt's, and herring sperm DNA should be stored at −20°C. Hybridization solutions are prepared fresh as required.

# Method

## *Restriction Endonuclease Digestion* (*See also* Chapter 31)

1. The amount of DNA to be digested will depend on the complexity of the source. A few nanograms of DNA from a recombinant molecule or virus will be sufficient. If single copy sequences are to be analyzed in genomic DNA from higher eukaryotes, 5–10 μg of DNA should be digested.
2. Prepare a digestion mixture containing: DNA sample, 0.1 vol. of 10× restriction enzyme buffer, 100 μg/mL nuclease free BSA, 2–3 units of restriction enzyme/μg DNA. (The final DNA concentration is usually 0.2–0.3 mg/mL). Incubate at 37°C (*see* Note 5) for 1–2 h (simple DNA), or 6–15 h (genomic DNA).

3. Stop the reaction by placing the tubes on ice, and adding 0.1 vol. 0.1*M* EDTA (pH 7.0).
4. If the samples were genomic DNA, the completeness of digestion should be checked by electrophoresis of an aliquot of the digest before proceeding (*see* Note 13).

## Agarose Gel Electrophoresis

(*See also* Chapter 7)

1. Prepare a 0.6–1.0% agarose gel by adding agarose powder to electrophoresis buffer and boiling until the solution is clear. Electrophoresis-grade agarose (e.g., Seakem from Marine Colloids Inc.) should be used. Ethidium bromide can be added to the molten gel to a final concentration of 1 μg/mL. Pour the agarose into the gel mold, insert the well-former ("comb") and allow to set for about 1 h. The concentration of agarose used will depend on the size of DNA fragments that are to be resolved (*see* Chapter 7).
2. Add 0.1 vol. loading buffer to the samples. A sample containing molecular weight marker DNA (e.g., λ digested with Hind III) should also be prepared. This can be radiolabeled with $^{32}P$ using polynucleotide kinase (Chapter 39), so that the marker bands will appear on the final autoradiograph.
3. Load the samples onto the gel and electrophorese at constant voltage until the Orange G has migrated to the end of the gel. Resolution of large DNA fragments can be optimized by running the gel overnight at a low voltage (about 1.5 V/cm).
4. If an unlabeled molecular weight marker has been used, photograph the gel on a UV source with a ruler positioned alongside the gel.

## Transfer of DNA

The DNA is now denatured in the gel and "blotted" out onto a nitrocellulose filter. Single-stranded DNA will bind tightly to the nitrocellulose, which can then be incubated with radiolabeled probes for specific sequences. Large DNA fragments (>10 kb) transfer very slowly, and should be broken down in the gel by partial depurination with dilute acid before transfer (3).

1. Place the gel in 0.25*M* HCl for 10 min.
2. Rinse the gel with distilled water, and place in denaturing solution, with gentle shaking, for 1–2 h.
3. Rinse the gel and place in neutralization solution for 1 h.
4. Set up the transfer as detailed in steps 5–16 below and illustrated in Fig. 2. The dimensions given are for a 20 × 20 cm gel.
5. Cut a square (24 × 24 cm) of Whatman No. 1 filter paper and fold over a 20 × 20 cm glass plate. Cut out the corners of the paper so that the edges can be folded down to act as a wick.
6. Place the plate and wick on supports (e.g., counting vials) in a glass or plastic dish, and pour 20 × 20 SSC into the dish to a level 2–3 cm below the plate.
7. Slide the gel onto the plate so that no air bubbles are trapped between it and the filter paper.
8. Cut a 20 × 20 cm sheet of nitrocellulose (Schleicher & Schuell, BA 85, pore size 0.45 μm, *see* Note 6). Always handle nitrocellulose with gloves that have been washed to remove powder. Wet the nitrocellulose by

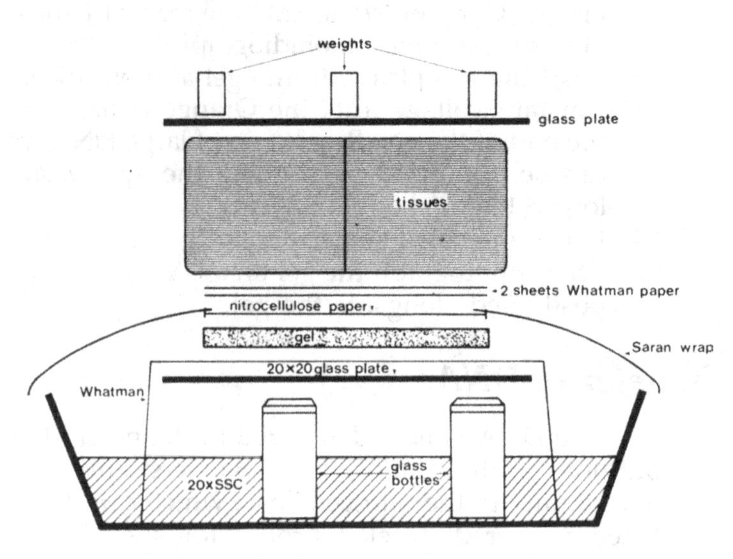


Fig. 2. Southern transfer setup.

flotation on 2 × SSC, and place it on the gel. Smooth out air bubbles trapped between the gel and filter.

9. Drape strips of cling film from the edges of the gel to the edges of the tray. This prevents evaporation of the SSC during transfer, and forces it to move through the gel.
10. Cover the nitrocellulose with two pieces of filter paper (20 × 20 cm) that have been wet in 2 × SSC.
11. Divide a box of tissues or paper hand towels in two, and place over the filter paper.
12. Put a glass plate on top of the tissues, followed by a 0.5–1 kg weight.
13. Leave the transfer at 4°C for 15–40 h.
14. After transfer, remove the tissues and filter paper. Cut the nitrocellulose into strips of dimensions just smaller than those of the hybridization chamber. Mark the position of the sample wells on the nitrocellulose and label each filter strip.
15. Soak the filter in 2 × SSC for 10 min, then bake them at 80°C for at least 2 h (*see* Note 7).
16. Filters may be stored at 4°C for several months before hybridization.

## Hybridization

The filters are now incubated with a $^{32}P$-labeled sequence-specific probe. The probe is usually labeled by nick translation (*see* Chapter 38), and will associate with its complementary sequence on the filter. Filters are coated with Denhardt's solution (4) and heterologous DNA before hybridization, to prevent nonspecific binding of the probe. Factors affecting the rate of hybridization have been reviewed (2,5). Conditions such as salt concentration and temperature are chosen to encourage hybridization (6). The filters are then washed in stringent conditions (low salt concentration) so that the probe will remain bound to only highly homologous sequences.

1. Wet the filters by floatation on 3 × SSC.
2. The filters are now prepared for hybridization by incubating them at 65°C, with shaking, in 50 mL of each of the following pre-heated solutions:

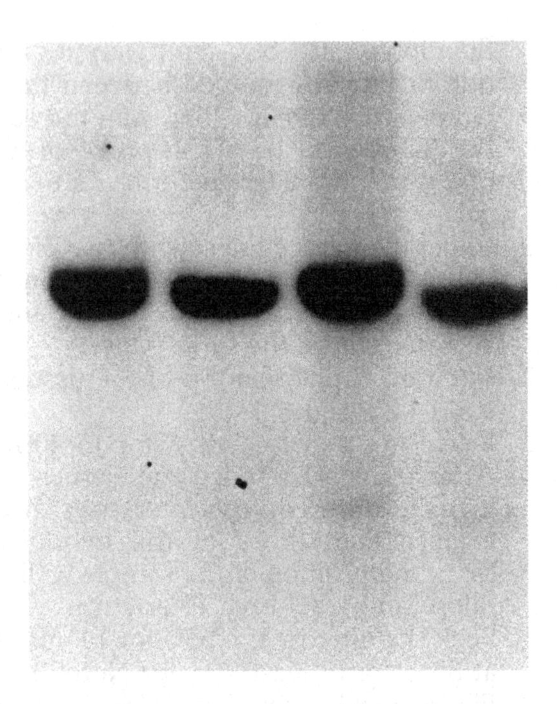

Fig. 3. Human DNA digested with the restriction enzyme Hpa 1, blotted, and hybridized with a β-globin cDNA probe. Only the 7.6 kb fragment that contains the β-globin gene is detected.

(a) 3 × SSC, for 30 min.
(b) 3 × SSC, 10× Denhardt's, for 60 min.
(c) Pre-hybridization solution for 30 min.

3. A 0.5 μg quantity of the sequence-specific probe is radio-labeled with $^{32}P$ by nick translation (*see* Chapter 38). If a single copy sequence is to be detected in genomic DNA, the specific activity of the probe should be at least $1 \times 10^8$ cpm/μg. Denature the $^{32}P$-labeled probe by heating in a boiling waterbath for 5 min. Cool on ice, and add to 10 mL of the prehybridization/hybridization solution in the hybridization chamber (*see* Note 8).
4. Transfer the filters to the hybridization chamber and incubate at 65°C, with shaking, for 24–48 h (*see* Note 9).

5. The filters are now given 6 × 50 mL washes in posthybridization wash solution at 65°C, with shaking, as follows: 4 × 1 min and 2 × 30 min.
6. Finally, wash the filters twice in 50 mL of stringent wash solution for 30 min at 65°C.

### *Autoradiography*

An X-ray film is now placed in contact with the filters and exposed. Sensitivity of detection is greatly enhanced by use of an intensifying screen (*see* ref. 2 for discussion).

1. Rinse the filters in 3 × SSC and reassemble, while moist, in a plastic bag.
2. Place the bag in a cassette, followed by the X-ray film and intensifying screen. Expose at −70°C for 1–4 d (*see* Note 10).
3. Develop and fix film according to the manufacturer's instructions.

## Notes

1. If a hybridization chamber cannot be obtained, the hybridization can be carried out in a sealed plastic bag. If a bag is used, ensure that air bubbles are excluded, and that the bag is properly sealed. The advantage of using the bag is that the nitrocellulose need not be cut into strips. However, the hybridization chamber produces "cleaner" backgrounds.
2. Most of the commonly used restriction enzymes are commercially available. They can be purified in the laboratory, but this would only be cost effective if large quantities are to be used.
3. Herring sperm DNA is used as a nonhomologous DNA that will saturate unused binding sites on the nitrocellulose. If DNA from a species of fish were being probed, it would then, of course, be necessary to use DNA from an unrelated organism at this step.
4. The salt concentration of the final wash will depend on the degree of homology between the probe se-

quence and the sample DNA. If, for example, a human probe is hybridized to human DNA, then a low salt concentration (high stringency) such as 0.1 × SSC would be used. At low stringencies (e.g., 1–2 × SSC), sequences of lesser homology (e.g., from other members of a multigene family or from different species) will be detected.

5. Most restriction enzyme digestions are done at 37°C, but some enzymes (e.g., Taq 1) have very different temperature optima (*see* Chapter 31).
6. Schleicher and Schuell nitrocellulose is widely used for binding DNA after transfer. Other filters or papers are commercially available, but these should be tested in controlled experiments before being used routinely. If DNA fragments of less than about 500 base pairs are to be detected, a chemically activated paper such as DBM paper (3) should be used, since the nitrocellulose does not bind small DNA fragments efficiently.
7. Nitrocellulose should be baked in a vacuum oven as a precaution. However, an ordinary oven can be used, provided that the temperature does not exceed 80°C.
8. Formamide can be included in the hybridization solution (5). This lowers the $T_m$ of the DNA, so that hybridization can be carried out at a lower temperature. However, formamide is expensive, toxic, and unnecessary.
9. A 24-h hybridization is sufficient for most purposes. This can be extended to 48 h if genomic DNA is being analyzed with a probe of low specific activity ($5 \times 10^7$–$1 \times 10^8$ cpm/μg). The rate of hybridization can be increased by the addition of dextran sulfate to the hybridization solution (3), but this is generally not necessary, and can cause intermittent high background signals (5).
10. The time required for adequate exposure of the autoradiograph will depend on the specific activity of the probe and the nature of the DNA being probed. Single copy sequences in genomic DNA should be detectable after a 1–2 d exposure with a probe of specific activity $1 \times 10^8$ cpm/μg or greater. Sensitivity of detection is greater at −70°C than at −20°C.

11. Used filters can be rehybridized to a second probe after removal of the original probe with NaOH. Soak filters in denaturing solution for 5 min, neutralization buffer for 2 h, and finally 3 × SSC for 15 min. Bake and prehybridize in the usual way.
12. The Southern transfer procedure is rather lengthy, taking about 5–8 d from restriction digestion to development of the autoradiograph. However, much of the time is "passive," e.g., leaving filters to incubate. No highly specialized equipment is required, but the cost of restriction enzymes and $^{32}P$-labeled nucleotide is considerable.
13. Incomplete digestion of DNA is a common problem, particularly in the case of genomic DNA. The degree of digestion may be monitored by removing an aliquot from the digest, adding 1 μg of λ-DNA to it, and incubating this in parallel with the original digest. If the expected pattern of λ-DNA fragments is not obtained, the digestion should be extended or repeated. The presence of spurious high molecular weight restriction fragments on the autoradiograph is an indication of partial digestion.
14. If the efficiency of transfer of the DNA out of the gel is poor, expected high molecular weight fragments may not be detected on the autoradiograph. The efficiency of transfer can be checked by restaining the gel after transfer.
15. A high background signal along the tracks of DNA suggests either that the probe contains repeat sequences or that the stringency of the final wash solution is too low.
16. If a high background that is randomly distributed over the filters is obtained, the final high stringency washes should be repeated. If this fails to remove the background, the filters can be treated with NaOH and rehybridized (*see* Note 11). Filters should always be handled with gloves. Once the filters hae been in contact with the radioactive probe, they should not be allowed to dry out until after the final stringency washes.
17. If no bands are detected on the autoradiograph, the following control experiments can be done:

(i) Blot 5–10 × $10^3$ cpm of Hind III digested λ-DNA onto filters and autoradiograph. If no bands are detected after an overnight exposure, remake 20 × SSC solution and use a different batch of nitrocellulose. Restain gel to check transfer.

(ii) Check the sensitivity of detection by loading 20 pg of probe DNA on the gel in addition to the samples. If only the probe is detected, check the recombinant plasmid for the presence of an insert. If neither probe nor samples are detected, check that the specific activity of the probe is at least 1 × $10^8$ cpm/μg, and prepare fresh hybridization and wash solutions.

## References


1. Southern, E. M. (1975) Detection of specific sequences among DNA fragments separated by gel electrtophoresis. *J. Mol. Biol.* **98,** 503–517.
2. Mathew, C. G. P. (1983) Detection of specific DNA sequences—the Southern blot, in *Techniques in Molecular Biology* (ed. Walker, J. M., and Gaastra, W.,) pp. 274–285. Croom Helm, London and Canberra.
3. Wahl, G. M., Stern, M., and Stark, G. R. (1979) Efficient transfer of large DNA fragments from agarose gels to DBM paper and rapid hybridization using dextran sulfate. *Proc. Natl. Acad. Sci. USA* **76,** 3683–3687.
4. Denhardt, D. T. (1966) A membrane-filter technique for the detection of complementary DNA. *Biochem. Biiophys. Res. Comm.* **23,** 641–646.
5. Maniatis, T., Fritsch, E. F., and Sambrook, J. (1982) *Molecular Cloning: A laboratory manual.* Cold Spring Harbor Laboratory, New York.
6. Jeffreys, A. J., and Flavell, R. A. (1977) A physical map of the DNA regions flanking the rabbit β globin gene. *Cell* **12,** 429–439.

# Chapter 10

# The Extraction and Isolation of DNA from Gels

*Wim Gaastra and Per Linå Jørgensen*

*Department of Microbiology, The Technical University of Denmark, Lyngby, Denmark*

## Introduction

As will be evident from a number of the following chapters (i.e., Chapters 31, 38–41, 51–53), gel electrophoresis of DNA is a widely used technique in molecular biology. In a number of cases, e.g., for such procedures as cloning and DNA sequencing, it is not sufficient just to analyze the DNA on these gels; the DNA must also be recovered from the gel. It is clear that the DNA in these cases has to be recovered in as high yields as possible and that the molecules should not be damaged. There are many published procedures for extracting DNA fragments from agarose or acrylamide gels (1–4), but none are very satisfactory. As mentioned in Chapters 38–41, agarose inhibits a number of enzymes used for labeling DNA molecules, for restriction, and for ligation. Acrylamide does not seem to inhibit most enzymes, but interferes with the electron microscopy of DNA. The procedures described below have all been used in our laboratory, albeit with varying degrees of success. The fact that a number of methods have not been included in this chapter does not mean that the particular method could not be of any use, but only that the authors are not familiar with it. The first step in each method is to locate the band of interest, either by staining the DNA with ethidium bromide or, if the DNA is radioactively labeled, by identifying by autoradiography, both of which are described elsewhere in this book and are therefore omitted from this chapter.

## Materials

### *Method 1*

1. Electrophoresis buffer: 5 m*M* Tris-acetate, pH ;8.0, or 5 m*M* Tris-borate-EDTA buffer, pH 8.0.
2. Sterilized dialysis bags with a cutoff of 3500 or 10,000 daltons, depending on the molecular weight of the DNA to be eluted.
3. 2-Butanol.
4. Redistilled phenol equilibrated with TE buffer. TE buffer: 10 m*M* Tris-HC1, pH 8, 1 m*M* EDTA (sodium salt).

### *Method 2*

1. Electrophoresis buffer: 6.8 m*M* Tris, 1.1 m*M* citric acid, 0.2 m*M* EDTA (sodium salt), at pH 8.1.
2. ISCO Model 1750 Electrophoretic Concentrator.

### *Method 3*

1. Eppendorf tubes.
2. Sterile cotton wool.

### *Method 4*

1. Elution buffer: 0.5*M* ammonium acetate, 10 m*M* magnesium acetate, 0.1% sodium dodecyl sulfate (SDS), 0.1

m*M* EDTA (sodium salt). The chemicals are dissolved in distilled water and the pH is not adjusted.
2. 96% Ethanol.

### Method 5

1. Whatman 3MM paper.

### Method 6

1. DEAE (diethylaminoethyl) paper (Schleicher and Schüll).
2. 1.5*M* NaCl and 1 m*M* EDTA (sodium salt), solution in $H_2O$.
3. Isopropanol.

### Method 7

1. Low melting agarose (Sigma).
2. TE buffer.
3. Redistilled phenol, equilibrated with TE buffer.

## Methods

### Method 1: Electroelution

1. A gel piece that contains the DNA fragment of interest is cut out of the gel and put into a dialysis bag, without damaging it, and 1–2 mL of the electrophoresis buffer is added.
2. The dialysis bags are placed in an electrophoresis tank of approximately 1 × 2 dm. The tank should have electrodes on the shorter sides. The bag should be parallel to these electrodes. Add enough electrophoresis buffer to cover the dialysis bag, usually to 6–7 mm height.
3. Electroelute the DNA at 150 V for approximately 45 min. If the gel has been stained with ethidium bromide, the duration of the electrophoresis can be determined by observing the elution of DNA using a long wave UV light.
4. After electroelution, reverse the polarity of cathode and anode for 0.5–1 min.

5. Carefully remove the buffer from the dialysis bag, without damaging the gel piece. If the recovery of the DNA is not complete, repeat steps 2–4.
6. Concentrate the DNA solution by extraction of water with 2-butanol. This procedure also removes any remaining ethidium bromide. The procedure for 2-butanol extraction of water from DNA solutions is as follows:

    (a) Add 1.5–1.8 vol of 2-butanol and mix for 15–20 s.
    (b) Separate the phases in an Eppendorf centrifuge for 2 min.
    (c) Discard the 2-butanol phase and repeat points (a) and (b) until a suitable volume of the lower phase is achieved.

7. Extract once or twice with phenol (e.g. *see* Chapters 39–41) to remove any agarose in solution, then extract the DNA solutions with ether, precipitate, and wash with ethanol, as described in Chapters 39–41.

## *Method 2: Electroelution*

1. A gel piece that contains the DNA fragment of interest is cut out of the gel and put into the big chamber of the sample cup of an ISCO Model 1750 Electrophoretic concentrator (*see* Fig. 1).
2. The electrophoresis tank and the sample cups are filled with electrophoresis buffer, to which 0.3 mg/L ethidium bromide is added.
3. Electroelute the DNA for 4–6 h at 3 W (4–6 mA). After the electroelution, the DNA is concentrated (sometimes even precipitated) on the dialysis membrane on the side of the anode. Because of the ethidium bromide, the DNA is readily visible in UV light.
4. Empty the sample cup apart from the last 200 μL above the small dialysis membrane. Resuspend the DNA in the last 200 μL by sucking up this 200 μL several times with a Gilson pipet, the tip of which is protected with a small piece of silicon tubing. The latter is to prevent puncturing the membrane and thereby losing the DNA solution.

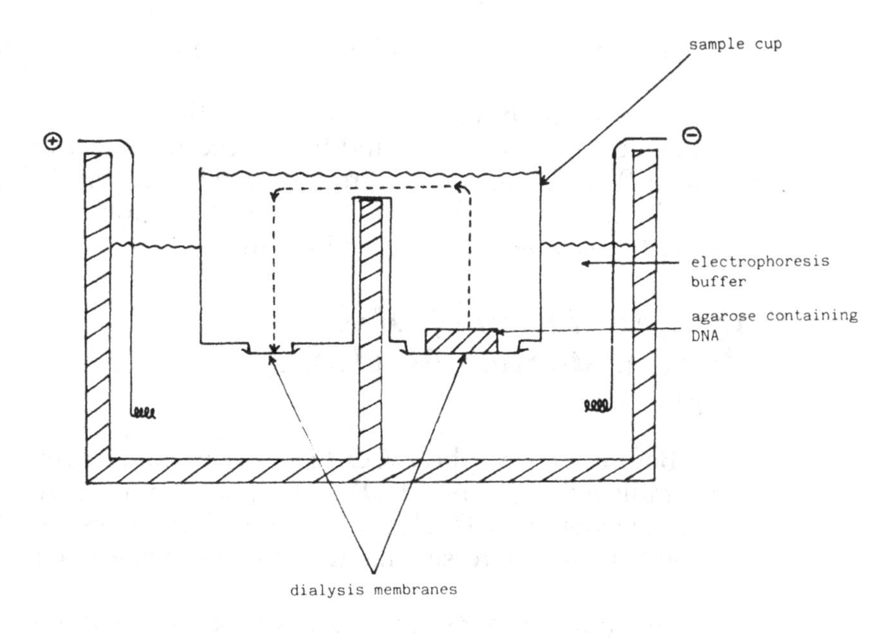


Fig. 1. Diagrammatic representation of the recovery of DNA from agarose gel by electroelution (Method 2).

5. To remove the ethidium bromide, extract the DNA with phenol and precipitate with ethanol, as described in Method 1.

## Method 3: Freeze–Squeeze Method

1. Make a small hole in the bottom of a 1.5 mL Eppendorf tube and place a small piece of sterile cotton wool on the bottom of the tube, covering the hole.
2. Put the tube with the cotton wool in another Eppendorf tube (3 mL).
3. Cut out the gel piece, which contains the DNA fragment of interest from the agarose gel, and put it in the upper Eppendorf tube, then place the whole construction in a −20°C freezer for 2 h, thereby destroying the structure of the gel.
4. Centrifuge the construction of the two tubes for 5 min in an Eppendorf centrifuge. The buffer from the gel,

containing the DNA, will be transferred to the bottom tube.
5. If not enough DNA is recovered, the gel piece is swollen again in buffer and the procedure is repeated.
6. The DNA is precipitated from the gel buffer with ethanol and is ready for further use. It may, however, be further cleaned-up as described in Method 1.

## *Method 4: Elution of DNA Fragments from Acrylamide Gel*

Before we had the ISCO Electrophoretic Concentrator, radioactively labeled DNA fragments for DNA sequencing were usually eluted from the 5% acrylamide gels on which they were separated in the following way:

1. After autoradiography, the DNA band of interest is sliced out of the gel, put into an Eppendorf tube, and the slice homogenized with a glass rod or another sharp object.
2. Add 600 μL of elution buffer, elute the DNA overnight at 45°C, then centrifuge the Eppendorf tube for 5 min.
3. Remove as much as possible of the supernatant, then filter the supernatant through siliconized glass wool to remove any gel debris. The glass wool is conveniently applied in a 200 μL pipet tip and the supernatant is forced through the glass wool with the help of a small pipeting balloon that is placed over the pipet tip.
4. Precipitate the DNA with 6 vol of cold ethanol (−70°C) and resuspend as needed for further use. Further cleaning-up can be carried out as described in Method 1.

## *Method 5: Electrophoresis of DNA into Whatman Filter Paper (5)*

1. Cut a piece of Whatman 3 MM paper and a piece of dialysis membrane, slightly larger than the size of the DNA band to be recovered. Wet the filter paper with

electrophoresis buffer and place it on the dialysis membrane.

2. Cut a slit in the agarose in front of the DNA band of interest and insert the filter paper. The filter paper should be inserted into the slit in such a way that it reaches the bottom of the gel and is backed by the dialysis membrane. The dialysis membrane should continue a little under the gel in the direction of the DNA.
3. Continue electrophoresis until all the DNA has migrated into the filter paper as determined under UV light. The DNA cannot move further because of the dialysis membrane. Remove the filter and dialysis membrane from the gel.
4. Place the filter paper in a 1.5 mL Eppendorf tube which has been prepared as described under Method 3, and recover the DNA containing gel buffer from the filter paper by centrifugation in the same way as described under Method 3.
5. Wash the dialysis membrane and the filter paper with the SDS containing buffer of Procedure 4 and collect the DNA solution again by centrifugation. This last step can be repeated several times to increase the yield.
6. Finally, precipitate the DNA and redissolve in the appropriate buffer. Alternatively, carry out further cleaning-up as described for Method 1.

## *Method 6: Binding of DNA to DEAE Paper (2)*

This method is essentially the same as described under Method 5, with the following exceptions. Instead of Whatman 3 MM filter paper, DEAE paper to which the DNA is electrostatically bound is used. The solution with which the paper is eluted is also different.

1. Cut a piece of DEAE paper (Schleicher and Schüll), slightly larger than the size of the DNA band to be recovered. Wet the paper and place it in a slit cut in the gel, in front of the DNA band of interest. Make sure that the paper reaches the bottom of the gel.
2. Continue electrophoresis until all the DNA has been bound to the DEAE paper. This can easily be moni-

tored under UV light. Remove the DEAE paper from the slit in the gel and put it in an Eppendorf tube.

3. Wash the paper once with TE buffer (for TE buffer, *see* Materials, Method 1), then elute the DNA from the DEAE paper with 600 μL of 1.5*M* NaCl, 1 m*M* EDTA solution, by incubation for 15–30 min at 65°C.
4. Spin the paper to the bottom of the Eppendorf tube and remove the supernatant as quantitatively as possible.
5. Wash the paper with another 600 μL of the above-mentioned salt solution, then combine the two supernatants and remove the ethidium bromide and precipitate the DNA with one volume of isopropanol.
6. Dissolve the DNA pellet after precipitation in the buffer needed next.

## *Method 7: Recovery of DNA from Low Melting Agarose*

1. Prepare an agarose gel as normal (Chapter 7) from low melting agarose, and run the gel in the cold room or in a refrigerator. Take care that the temperature of the electrophoresis buffer remains below 10°C.
2. After electrophoresis cut out the bands that have been visualized with ethidium bromide, then place the agarose blocks in Eppendorf tubes and add one time the agarose volume of TE buffer. Incubate for 10 min at 65°C.
3. Quickly add one volume of phenol, mix, centrifuge in an Eppendorf centrifuge, and remove the aqueous layer. Repeat this step twice.
4. Remove any remaining phenol by three extractions with ether, then precipitate the DNA twice with 96% ethanol.
5. Dissolve the DNA pellet in the desired buffer for your next step.

## Notes

1. The yields obtained with the various methods lie between 50 and 90%. This, of course, depends largely on

the size of the DNA fragment to be recovered. Usually the higher yields are obtained with the electrophoretic methods. Lower yields are obtained with methods that depend on diffusion of the DNA out of the gels.

2. As mentioned in the introduction, DNA solutions recovered from acrylamide or agarose gels usually contain some of the gel material from which the DNA was recovered. Since these contaminants from gel debris may interfere with subsequent enzymatic reactions, it could be desirable to clean up the DNA solution afterwards. A number of methods, such as phenol extraction, chromotography on DEAE cellulose or hydroxyapatite, and density equilibrium have been described for this purpose (2), but are not further discussed here. However, the procedure described at the end of Method 1 is generally suitable.
3. Although very high yields have been obtained with the ISCO sample concentrator, it has, of course, the disadvantage of being rather expensive in comparison with the other methods. The same holds for the method employing the low melting agarose, which is also rather expensive.
4. We have observed that during the various extraction procedures of DNA, DNAses are easily introduced into the system. It is therefore advisible to wear gloves and use sterilized materials and buffers while handling and extracting gel pieces.
5. If the apparatus is available, it is usually very helpful to take a polaroid picture, before and after the DNA bands have been cut out correctly.

## References


1. Yang, R. C. A., Lis, J., and Wu, R. (1979) Elution of DNA from agarose gels after electrophoresis. *Meth. Enzymol.* **60,** 176–182.
2. Smith, H. O. (1980) Recovery of DNA from gels. *Meth. Enzymol.* **65,** 371–380.
3. Chen, C. H., and Thomas, Jr., C. A. (1980) Recovery of DNA segments from agarose gels. *Anal. Biochem.* **101,** 339–341.

4. Drelzen, G., Bellard, M., Sassone-Corsi, P., and Chambon, P. (1981) A reliable method for the recovery of DNA fragments from agarose and acrylamide gels. *Anal. Biochem.* **112,** 295–298.
5. Girvitz, S. C., Bacchetti, S., Rainbow, A. J., and Graham, F. L. (1980) A rapid and efficient procedure for the purification of DNA from agarose gel. *Anal. Biochem.* **106,** 492–496.

# Chapter 11

# One-Dimensional Electrophoresis of Nucleic Acids in Agarose Using Denaturation with Formaldehyde and Identification of $^{3}H$-Labeled RNA by Fluorography

***Theodore Gurney, Jr.***

*Department of Biology, University of Utah, Salt Lake City, Utah*

## Introduction

The procedure described in this chapter is used to display single-stranded nucleic acids according to their sizes, within the range of 0.5 to 30 kilobases (kb). Possible applications include examining products of in vitro synthesis, hybrid-selected RNAs from total cellular nucleic acids, and Northern blots. The method works equally well with RNA and DNA.

Full denaturation is needed to determine single-strand sizes unambiguously because partial hydrogen bond formation within or between polynucleotides will affect the eletrophoretic mobility (*1*). DNA may be denatured and electrophoresed in alkali, but RNA is hydrolyzed at a pH greater than 11.3, which is necessary in order to break all the hydrogen bonds. Continuous heat denaturation is not compatible with agarose, which must be used instead of polyacrylamide for larger polynucleotides. Fortunately, there are three denaturing agents, formaldehyde (*1*), glyoxal (*2*), and methyl mercuric hydroxide (*3*), that are compatible with both RNA and agarose. Each forms adducts with the amino groups of guanine and uracil after heat denaturation, thereby preventing hydrogen bond reformation at room temperature during electrophoresis.

Formaldehyde is less toxic, less expensive, and more stable than the other two, although it is quite dangerous. The US Occupational Safety and Health Administration registered formaldehyde as a weak carcinogen in 1982; therefore, all work with formaldehyde in open containers must be carried out in a fume hood. All wastes containing formaldehyde should be considered as hazardous in our environment and cannot be flushed into municipal sewers.

The methods described here are designed for radiolabeling procedures, either electrophoresis of radiolabeled nucleic acids or else hybridization after electrophoresis to radiolabeled probes, that is, Southern blots and Northern blots. Nucleic acids treated with formaldehyde and glyoxal will bind well to nitrocellulose used in blotting (*4,5*), and vacuum baking makes the bound nucleic acids hybridizable again. Radioactivity is detected by autoradiography on X-ray film.

In this article, I also describe a procedure used to detect $^3$H-labeled electrophoresed nucleic acids, based on published methods (*5,9,10*) (*see* also Vol. 1). The gel is impregnated with PPO, a fluorescent compound that emits light when bombarded with the $^3$H beta particles. Direct gel impregnation must be used because the $^{3}H$ beta particle is too weak to reach the film or the fluorescent screen in a medical X-ray film holder. PPO is soluble in methanol, but not in water. Hence the gel is dehydrated in methanol, soaked in a PPO–methanol solution, and then PPO is trapped in the gel by precipitation in water. The gel is then dried and exposed to X-ray film at −70°C. The cold temperature is required to produce the proper wavelength of fluorescent light (*9,10*). The method given here uses different gel mounting paper from that of our previous version (*5*) and is a distinct improvement.

Optical methods of detecting nucleic acids are also possible, but light scattering by agarose limits sensitivity and denatured nucleic acids stain weakly with ethidium bromide. Acridine orange staining (*2*) is probably the most sensitive optical procedure for use with agarose and formaldehyde.

# Materials

1. Formaldehyde. The common reagent-grade 37% (w/v) solution contains 10–15% methanol as a preservative. The methanol does not interfere with the procedures. Formaldehyde solutions should have little or no paraformaldehyde, seen as a visible precipitate. To prevent paraformaldehyde formation, formaldehyde solutions should be stored at temperatures above 20°C. Tris buffer must not be used with formaldehyde because of reaction with the amino group. Possible buffers used with formaldehyde are phosphate (*1*) and triethanolamine (*6*), which is used here.
2. Triethanolamine, practical grade or better. It is a viscous liquid.
3. Agarose, electrophoresis grade.
4. PPO (2,5-diphenyloxazole), scintillation grade.
5. Formamide, vacuum distilled or deionized. It is stored in quantities of 50 mL at −20°C.
6. Methanol, reagent grade.
7. Nucleic acid buffer for dissolving RNA and/or DNA from ethanol precipitates: 10 m*M* triethanolamine-HCl, diluted from 1*M*, pH 7.4; 1 m*M* EDTA, diluted

from 500 m$M$, pH 9.0; and 0.5% SDS diluted from 10% (w/v). The buffer and its component reagents are stored at room temperature. Solutions of nucleic acids in nucleic acid buffer are stored at −20°C.

8. The gel buffer in the gel, in the sample, and in the buffer reservoirs is: 20 m$M$ triethanolamine, pH 7.4; 2.5 m$M$ EDTA; and 2.2$M$ formaldehyde. The buffer is mixed as 5× concentrate: Weigh out 3 g of liquid triethanolamine into a 250 m$M$ beaker. Add 89 mL of 37% formaldehyde (**N.B.**: work in a hood), 5 mL of distilled water, and 2.5 mL of 0.5$M$ $Na_4EDTA$, pH 9. Adjust the pH from about 9 to 7.4 with 4$N$ HCl. Store the buffer tightly capped at room temperature in the hood. Prepare enough 1× gel buffer for the reservoirs of the electrophoresis apparatus. The reservoir buffer may be reused at least 10 times if the two reservoirs are mixed during the run or after the run. If you do not use buffer mixing during the run, the reservoirs should each hold at least 250 mL of the 1× buffer.
9. Sample preparation buffer is prepared just before use: Mix 10 volumes of formamide with 4 volumes of 5× gel buffer.
10. Mock sample is prepared just before use: Mix 3 parts nucleic acid buffer with 7 parts sample preparation buffer.
11. 20× SSC: 3$M$ NaCl, 0.3$M$ trisodium citrate, pH 7.0, and is stored at 22°C.
12. 2% Glycerol: 2% (v/v) in water, stored at 22°C.
13. The slab gel apparatus should be the flat-bed type used for Southern blots (*see* Chapters 7 and 9). The apparatus must allow removal of the unsupported agarose slab for processing. The size of the slab was chosen with 13 × 18 cm X-ray film in mind, with electrophoresis in the longer dimension. The apparatus has a comb with 23 teeth to cast 23 sample slots of 3 × 1.5 mm in area and 2 mm in depth across the width (12 cm) of the bed at one end. The flat bed of agarose is connected electrically to two 250-mL reservoirs of 1× gel buffer through agarose bridges 6 mm thick. Several designs of apparatus are satisfactory.
14. The power requirements are 30–60 V dc, constant voltage, adjustable and regulated. Current through a 3

mm thick, 12 cm wide gel is about 25 mA. A household appliance timer can be used to turn the power on and off.

15. A shaking apparatus is used in gel processing. The best shaking is back-and-forth, amplitude 2 cm, period 1 s. A second choice is circular shaking in a horizontal plane, radius 1 cm, period 1 s.
16. A gel dryer is used with timed 70°C heat.
17. Two vacuum sources are used, a water aspirator with a glass 200-mL trap, and a mechanical vacuum pump, 60 L/min, capable of 100 mtorr, with two glass cold traps in series. Both vacuum sources are necessary.
18. X-ray equipment includes the most sensitive X-ray film, a light-tight mounting press, a −70°C freezer, developing chemicals, and a Wratten 6B safelight filter. A satisfactory homemade press is two sheets of 5 mm-thick fiberboard, held together with spring steel binder clips and holding between them a light-tight envelope.
19. The Southern blotting apparatus and supplies are described in Chapter 9.
20. Two types of mounting paper are used, a heavy porous blotter paper, as used for mounting polyacrylamide gels, and a thinner paper that is strong when wet, such as artist's water-color paper.
21. Casein glue is used in gel mounting.
22. Bromphenol blue, 1% (w/v) in water is used as a tracking dye.
23. Plastic wrap, of the type used in food preparation, is used in gel processing.

# Methods

## *Concentrating Nucleic Acid Solutions*

The nucleic acids must be deproteinized and then concentrated solutions made in the triethanolamine nucleic acid buffer. A high concentration is required because the amount of solution used per sample is only 3 µL. If you are using radioactive nucleic acids, the radioactivity will determine the concentrations; for instance, 1000 cpm of $^{3}H$ or 100 cpm of $^{32}P$ in one electrophoretic species will make a band after overnight exposure (*see* Note 1). If you are preparing nonradioactive RNA for Northern blots, you should use at least 5 mg/mL RNA (Note 2).

1. Adjust the sodium ion concentration to at least 100 m*M* in your dilute nucleic acids.
2. If the concentration of nucleic acids is less than 20 μg/mL, add purified tRNA to 20 μg/mL.
3. Put 0.4 mL of adjusted solution in a 1.5 mL microfuge tube, add 1.0 mL (2.5 vol) of 95% ethanol, then mix and chill for at least 4 h at −20°C. Centrifuge (5 min, 8000*g*, 2°C) and then gently decant the supernatant; the pellet may be loose and will probably be invisible.
4. To aid redissolving, you should desalt the sample further. Fill the tube half full with 70% ethanol at 2°C, mix vigorously, and centrifuge again (1 min, 8000*g*, 2°C).
5. Decant or draw off the supernatant carefully, inverting the tube in the process. Keep the tube upside down while you wipe the inside walls with a tissue to get rid of traces of ethanol, or else you will resuspend the pellet in the residual ethanol. You can also vacuum-dry the tube to remove ethanol, but this is not necessary.
6. Redissolve the pellet in a small volume, e.g., 10 μL, of nucleic acid buffer at room temperature. Pipet the dissolving nucleic acids up and down about 50 times. If you are studying radioactive samples, determine the radioactivity at this point.

## Sample Preparation

Nucleic acids are heat-denatured in formamide plus formaldehyde. Formamide lowers the melting temperature and consequently lowers the risk of temperature-dependent nicking. Also formamide makes the sample dense enough to layer it in the sample slots without having to supplement it with sucrose or glycerol.

1. In a clean microfuge tube of 500 or 1500 μL capacity, mix 3 μL of the dissolved nucleic acids and 7 μL of sample preparation buffer. Mix by vortexing the capped tube very vigorously. Centrifuge briefly to collect the sample at the bottom of the tube. (Heavy formamide resists casual mixing.)
2. Incubate the samples at 55°C for 15 min. After incubation, the samples can be stored capped for several hours at room temperature.

## Gel Preparation and Electrophoresis

The agarose concentration must be between 2% (w/v), which is the solubility limit, and 0.5%, which begins to be too difficult to process. A solution of 0.7% agarose is the concentration to use in most applications because it resolves the widest range of polynucleotide sizes, from below 1 to above 25 kb. Agarose electrophoresis is not the method of choice for smaller molecules, however (*see* Fig. 1 and Note 3).

1. The agarose gel material, prepared as described below, is used for both the 3-mm flat-slab resolving gel and for thicker bridges connecting the slab electrically to buffer reservoirs. (Some designs do not use bridges, for example, the newer "submarine gels"; *see* Chapter 7). Determine the volume of agarose you will need. For example, my bridges require 50 mL apiece and the flat slab requires 80 mL. Depending on the apparatus design, you may have to pour the bridges first and let them harden before pouring the slab. The agar is melted in boiling water in the fume hood because of the presence of formaldehyde. The bridges may be poured several days ahead of time, but the slab should be poured immediately before use (Note 4). Before pouring the slab, be sure it is level. Use a carpenter's bubble level. If your apparatus comes with a bubble level, calibrate it once with another one.

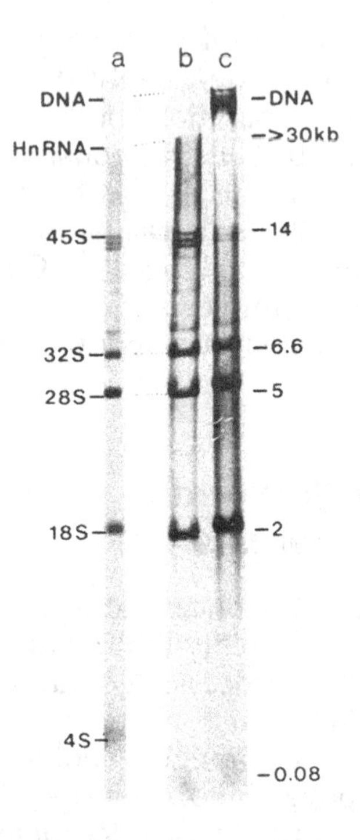


Fig. 1. Electrophoresed nucleic acids from mouse Balb 3T3 cells labeled 2 h in vivo with $^3$H-uridine. The cells were synchronized in the cell cycle by contact-inhibition followed by serum-stimulation, as described in ref. *11*. *Lane a:* G1-phase cells, 5 h after stimulation, 0.6% agarose, $1 \times 10^4$ cpm $^3$H, 2 d film exposure. The sample slot was 3 mm wide. Power was turned on 3 h after loading the sample. *Lane b:* G1-phase cells, 5 h after stimulation, 0.75% agarose, $2.5 \times 10^4$ cpm $^3$H, 3 d film exposure. The sample slot was 6 mm wide. Power was turned on immediately after loading the sample. *Lane c:* S-phase cells, 20 h after stimulation; other conditions were the same as Lane b.
Points to note: (1) the H-shaped bands in lanes b and c (compare with a) can be avoided by waiting before turning on the power. (2) Narrow samples lanes do not result in loss of resolution; resolution in the 14 kb region (45 S rRNA) is good enough to distinguish subspecies differing by 300 nucleotides.(3) 4S RNA is not well resolved although it is detected. (4) Labeled high mw nuclear DNA (present only in S-phase cells, Lane c) enters the gel slightly but does not occupy a region containing high mw RNA. Hence, in many experiments, nuclear DNA should not interfere with RNA analysis, as long as DNA is in solution and is of high molecular weight.

2. Next adjust the comb to cast the slots that will hold samples. The teeth of the comb should be held off the bottom of the slab by at least ½ mm. Be sure that you can pass a sheet of thick paper under the teeth. If you have a choice of comb tooth size, pick smaller ones, e.g., 3 mm in width, because you may then load more samples. In my hands, a 3 mm tooth makes a sample slot of about 10 μL.
3. Now you are ready to pour the slab, in the fume hood. The following is a recipe for 100 mL of 0.7% agarose, used for the bridges or the slab. In a 250 mL Ehrlenmeyer flask, mix 70 mg of agarose powder and 80 mL of water. Heat the suspension in boiling water for 5–10 min. Remove the flask from the boiling water and immediately swirl it to dissolve the agarose. Immediately add 20 mL of room-temperature 5× formaldehyde gel buffer while mixing. Mix thoroughly and pour the slab while the agarose is very hot. (You may have to let the agarose cool a little if the heat will damage the apparatus; consult your instruction manual.) Rinse the flask. Wait 30–60 min for the slab to cool and harden. The room temperature must be below 30°C for agarose to harden properly. Use a gentle rate of air flow in the fume hood, to avoid drying the agarose and to avoid making waves in it while it is hardening. Test a corner of the slab for hardness. After hardening, lift and then rinse the comb immediately.
4. Flood the top surface of the slab with an excess of 1× gel buffer and fill the sample slots. If your gel is not of the submarine design (under a layer of buffer), cover the slab with a smooth layer of plastic wrap. Eliminate bubbles between the plastic wrap and the gel, since a bubble over your sample will always make a streak (*see* Chapter 3, Fig. 1g).
5. Before loading the samples, connect electrodes from the power supply, and turn the power on to adjust the voltage. The negative electrode attaches near the sample slots and the positive electrode at the far end. For a slab 12 cm wide, 20 cm in length, 3 mm thick, 250 mL reservoirs, and no buffer recirculation, 650 V-h will move tRNA (the smallest nucleic acid) about 15 cm. Larger reservoirs or buffer recirculation gives

more extensive electrophoresis. You should keep current below 25 mA to avoid resistive heating, unless you have water-cooling on both sides of the slab. The heating therefore limits the possible voltages. In addition, you should wait 3 h after loading before applying power (see below). These considerations usually mean an overnight run. Choose your voltage, probably between 30 and 60 V, and check the ammeter for a complete circuit, so that a current flows in the gel. Then turn the power off, and set a timer to turn it on 3 h after loading the samples in the gel.

6. Peel back the plastic wrap just enough to expose the sample slots. Be sure that you can put the wrap back without bubbles. Mix approximately 0.1 μL of 1% bromphenol blue with the sample, and apply the sample to a sample slot filled with buffer. This takes a steady hand; you may wish to practice first by filling all unused sample slots with mock sample (Note 5). The sample slots of apparent size 3 × 1.5 × 2 mm will hold 10 μL (barely), because the miniscus effect between gel and comb actually makes the slots a little deeper than 2 mm. The lower limit of sample volume is determined only by the lower limit of accurate pipeting that is about 2 μL in my hands. After loading the samples, replace the plastic wrap, without bubbles. The new "submarine" designs of apparatus (Chapter 7) eliminate bridges, bubbles, and plastic wrap, but loading samples through a rather deep layer of buffer is more difficult.
7. After loading, the samples should rest for 3 h with no voltage applied, to allow formamide to diffuse (partially) out of the sample slots. Diffusion of small molecules makes a more uniform initial electric field across the samples. Otherwise, the formamide produces an inhomogeneous field and the electrophoretic bands will be H-shaped rather than flat. (Compare lanes a and b of Fig. 1.) The three-hour wait need not lengthen an overnight run because you can probably compensate by using a higher running voltage to give the same 650 volt hours by morning. Bromphenol blue dye should move 12–15 cm from the origin. (Note 6.)

## *Gel Processing for Blotting to Nitrocellulose* (4,7,8)

1. Put 100 mL of water in a dish and float a pre-cut sheet of nitrocellulose paper in the water. Get the blotting apparatus ready (*See* Chapter 9).
2. Remove the formaldehyde gel buffer from the electrophoresis apparatus. Pipeting with a 25 mL pipet works well enough. Save the buffer, capped. Remove excess buffer from the agarose slab with tissue paper and cut the gel free from adhering parts with a knife.
3. Pick up the part of the apparatus that holds the slab of agarose gel and hold it upside down 2–3 cm over the 3MM paper of the blotting apparatus. Pry a corner of the gel free to start separating the gel from the apparatus. The gel should drop, unbroken, onto the paper. Any alternate method of getting unattached agarose slab onto the 3MM is good enough. You should practice this part before committing your samples.
4. Spread the wet nitrocellulose sheet over the gel. Eliminate bubbles between the gel and the nitrocellulose sheet using gloved hands. From this point on, blotting is done by Southern's procedure (*8*) using 20x SSC as the transfer buffer (*see* Chapter 9).

## *Gel Processing for Detection of $^3$H-Labeled Nucleic Acids* (5,9,10)

1. Put 200 mL of 10% (v/v) glacial acetic acid in a 20 x 30 cm baking dish.
2. Remove the formaldehyde gel buffer from the electrophoresis apparatus with a pipet; save the buffer for reuse. Cut the slab free from the edges with a knife.
3. Pick up the slab-support and hold it upside-down 2–3 cm over the baking dish. Pry a corner of the gel free to allow the whole slab to drop unbroken into the acetic acid in the baking dish. Cover the dish with plastic wrap.
4. Shake the gel for 30 min. Stop the shaker immediately if the gel sticks to the dish and carefully detach the gel

from the dish. Remove the acetic acid into the hazardous waste, by pipeting or pouring.

5. Add 200 mL of methanol to the dish. Let the gel sit in the dish covered with plastic wrap, without shaking, for at least 15 min. The gel becomes especially sticky at this point and shaking might rip it. Then start to shake it, but stop if the gel sticks, and free it from the dish. Shake for at least 30 min.
6. Remove the methanol by pouring or pipeting into the hazardous waste and add 200 mL of fresh methanol. Shake the gel, covered, for at least 45 min.
7. While the gel is in its second methanol rinse, prepare 100 mL of 16% (w/w) PPO in methanol. Warm it gently (approx. 50°C, no flame) to dissolve, and keep it at 25–30°C.
8. After the second methanol rinse, remove the rinse as above and transfer the gel to a sheet of dry heavy blotter paper by pressing the paper against the gel while turning the dish upside down. Place the supported gel on the porous metal screen of the gel dryer with the paper in contact with the metal. Then cover the gel, first with thin plastic wrap, next with a stiffer sheet of plastic (supplied with commercial dryers), and finally with a silicone rubber sheet vacuum seal. Attach the water aspirator and apply vacuum, but no heat. Turn the vacuum off after a minute or when nearly all of the methanol has been drawn into the trap. Release the vacuum by lifting the rubber flap. Avoid drawing water from the aspirator into the gel dryer (Note 7).
9. Dry the baking dish with a towel and put the 16% PPO into it.
10. The compressed gel should be very thin and should be (barely) strong enough to be lifted unsupported. It should be still wet with methanol. Lift the gel quickly (to avoid rehydration from water in the atmosphere) from the dryer to the PPO solution in the baking dish. Cover the dish and shake for 30–60 min at 25–30°C. A temperature of at least 25°C is required to keep the PPO in solution.

11. Put 300 mL of distilled water into another baking dish and another 300 mL of water in a beaker. Pick the gel out of the PPO-methanol with two gloved hands and lay it on the water in the dish. It should float. Immediately pour water in the beaker over the gel. The gel should become uniformly white and should sink. Leave it there while you attend to the next three steps.
12. Pour about 50 mL of 2% glycerol (in water) into another baking dish.
13. Cut three thicknesses of paper towel and one piece of plastic wrap to a size between those of the gel and the gel dryer. Place the towels on the metal screen of the dryer.
14. Cut a piece of water-color paper to a size slightly larger than the gel. Wet one side, then the other, with 2% glycerol. Lay the wet paper on a clean patch of lab bench. Wet the bench on one side of the paper, then spread casein glue on the watercolor paper, 0.5–1 mL/300 $cm^2$ to make sticky paper.
15. Pick the gel out of the water with two gloved hands by one edge. Let it drain for 3 s, then drag the opposite edge along the wetted lab bench toward, and then onto, the sticky paper. Flop the gel down on the sticky paper. It is important to get no glue on the top side of the gel since it blocks fluorescence. Keep the glue off your gloves. You should probably practice this step before committing valuable samples.
16. Place the paper plus gel on the towels on the gel dryer. Cover with plastic wrap. Rub the gel gently through the plastic wrap to squeeze out bubbles and to get a good bond of the gel to the paper. Cover with the gel dryer's stiff plastic sheet and then the rubber vacuum seal. Dry the gel with high vacuum plus heat for 30 min. Use the mechanical pump and two cold traps.
17. While the gel is drying, you may recycle the PPO. The excess PPO may be recovered by precipitation in water, which is simply the mixing of the contents of all the baking dishes and recovering the precipitate by filtration (*See also* Chapter 17 of Vol. 1). Rinse the

dishes with methanol to remove traces of remaining PPO; it cannot be removed by usual washing procedures (Note 8).

18. Remove the gel from the dryer, trim away extra paper with scissors and mount the gel in the press with X-ray film. Expose the film at −70°C. The development of the film is described in Chapter 17 of Vol. 1.

## Notes

1. Radioactivity of concentrated nucleic acids can be determined by spotting 1 or 2 μL on a small piece of Whatman GF/C paper, vacuum drying the paper, and counting in a toluene-based scintillation fluid, without a solubilizer. Only glass fiber paper can be used with $^3H$.
2. The concentrations of nucleic acids from whole mammalian cells can be estimated approximately as 10 pg DNA and 20 pg RNA per cell.
3. Your first experience with agarose gels may be frustrating if you are used to polyacrylamide gel electrophoresis. The agarose gels have no elasticity and next to no tensile strength. You should handle the gel with support at all times. There is no way to lift a 0.7% agarose gel unsupported without ripping it. If you do rip a gel, it may be pieced together like a jigsaw puzzle on the sticky paper support just before the final drying.
4. Both the 1x gel running buffer and the agarose bridges may be reused several times if the buffers at the opposite ends of the gel are mixed together between uses, to re-equilibrate ions displaced by electrophoresis.
5. The reason for filling the unused sample slots with mock sample is to make the electric field uniform near the unused slots during electrophoresis.
6. Transfer RNA and 5 S RNA run about 10% *ahead* of the dye during electrophoresis.
7. The first of the two gel drying steps, removal of the methanol and the concomitant compression of the 3 mm thickness to a 0.5 mm thickness, must be done

with a vacuum source not ruined by methanol. This step generates 80–100 mL of methanol, which can go right through dry-ice cold traps into an expensive mechanical vacuum pump.

8. If you do fluorography with ethidium bromide or acridine orange, you should know that PPO is also a strongly fluorescent compound, and that a little PPO contamination can ruin other fluorography. It would be best to use separate glassware and gloves with PPO. PPO fluorescence is yellow-green, to distinguish it from ethidium bromide.

## Acknowledgments

I thank Elizabeth Gurney, Chris Simonsen, Arnold Oliphant, Dean Sorenson, and Paul Hugens for help and several insights. This work was supported by USPHS Grant GM 26137 and a grant from the University of Utah Research Committee.

## References

1. Lehrach, H., Diamond, D., Wozney, J. M. and Boedtker, H. (1977) RNA molecular weight determinations by gel electrophoresis under denaturing conditions, a critical reexamination. *Biochemistry* **16,** 4743–4751.
2. McMaster, G. K., and Carmichael, G. G. (1977) Analysis of single and double stranded nucleic acids on polyacrylamide and agarose gels by using glyoxal and acridine orange. *Proc. Natl. Acad. Sci. USA* **74,** 4835–4838.
3. Bailey, J. M., and Davidson, N. (1976) Methylmercury as a reversible denaturing agent for agarose gel electrophoresis. *Anal. Biochem.* **70,** 75–85.
4. Thomas, P. S. (1980) Hybridization of denatured RNA and small DNA fragments transferred to nitrocellulose. *Proc. Natl. Acad. Sci. USA* **77,** 5201–5205.
5. Gurney, T., Jr., Sorenson, D. S., Gurney, E. G., and Wills, N. M. (1982) SV40 RNA: Filter hybridization for rapid isolation and characterization of rare RNAs. *Anal. Biochem.* **125,** 80–90.
6. Lizardi, P. M. (1976) The size of pulse-labeled fibroin messenger RNA *Cell* **7,** 239–245.

7. Goldberg, D. A. (1980) Isolation and partial characterization of the *Drosophila* alcohol dehydrogenase gene. *Proc. Natl. Acad. Sci. USA* **77,** 5794–5798.
8. Southern, E. M. (1975) Detection of specific sequences among DNA fragments separated by gel electrophoresis. *J. Mol. Biol.* **98,** 503–513.
9. Bonner, W. M., and Laskey, R. A. (1974) A film detection method for tritium-labeled proteins and nucleic acids in polyacrylamide gels. *Eur. J. Biochem.* **46,** 83–88.
10. Laskey, R. A., and Mills, A. D. (1975) Quantitative film detection of $^{3}H$ and $^{14}C$ in polyacrylamide gels by fluorography. *Eur. J. Biochem.* **56,** 335–341.
11. Foster, D. N., and Gurney, T. Jr. (1976) Nuclear location of mammalian DNA polymerase activities *J. Biol. Chem.* **251,** 7893–7898.

# Chapter 12

# Gel Electrophoresis of RNA in Agarose and Polyacrylamide Under Nondenaturing Conditions

**_R. McGookin_**

*Inveresk Research International Limited, Musselburgh, Scotland*

## Introduction

This article details two methods for separation and visualization of RNA under nondenaturing conditions, i.e., where the secondary structure of the molecules is left intact during electrophoresis. The first method describes electrophoresis in a 2% (w/v) agarose gel in a dilute, neutral phosphate buffer. The second deals with electrophoresis in a linear gradient of polyacrylamide based on the buffer system of Loening (*1*).

The methods differ sufficiently in the results they give to merit separate description here. The agarose gel system is quick and easy to perform, making it ideal for rapidly checking the integrity of RNA immediately after extraction before deciding whether to process it further. Electrophoresis may be finished in less than 1 h, the 18 and 28S rRNAs are clearly resolved and any degradation or DNA contamination is easily seen (Fig. 1). The polyacrylamide gel system is a linear gradient of 2.4–5% (w/v) with a 2% (w/v) spacer gel on top. Although it is slow to set up and run, the resolution normally observed is much greater. It

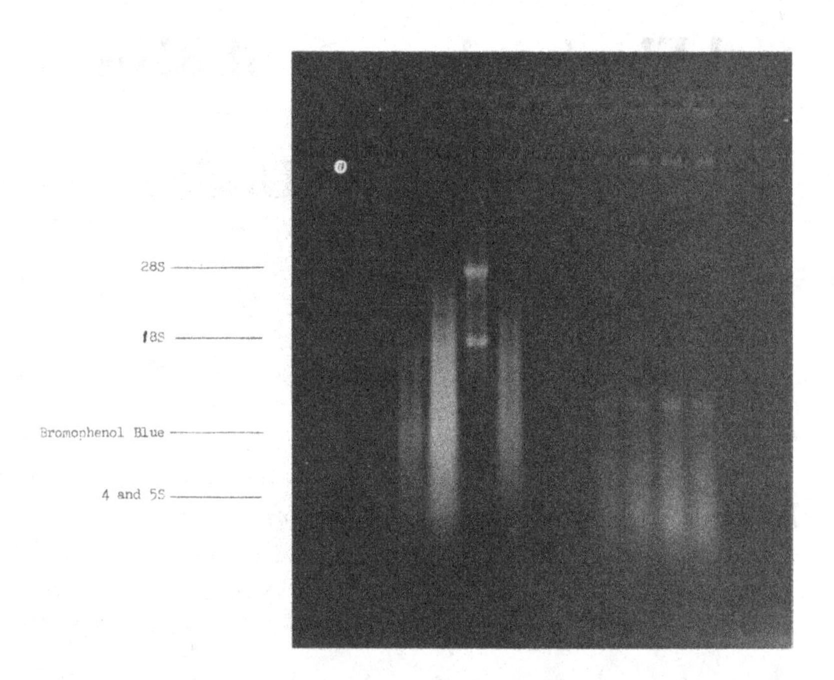


Fig. 1. Neutral phosphate gel of various RNA samples. The gel was a 2% (w/v) agarose horizontal gel run as described in the Methods. The wells are numbered from left to right.

| *Well No.* | *Sample* |
|---|---|
| 3 | Degraded *E. coli* RNA |
| 4 | Degraded *E. coli* RNA |
| 5 | 10 μg total cytoplasmic RNA from a human cell line |
| 6 | 10 μg of poly(A)$^+$ RNA from a human cell line |
| 9, 10, 11, 12 | Various concentrations of soluble material after a 2*M* LiCl precipitate of unbound RNA after oligo (dT)-cellulose chromatography |

is usual to clearly differentiate rRNA species from organelles and cytoplasm and to resolve tRNAs from 5S rRNA (Fig. 2). If, for example, one is interested in an abundant class of mRNA that is developmentally regulated, this system will provide the best chance of detecting such a species in nondenaturing gels.

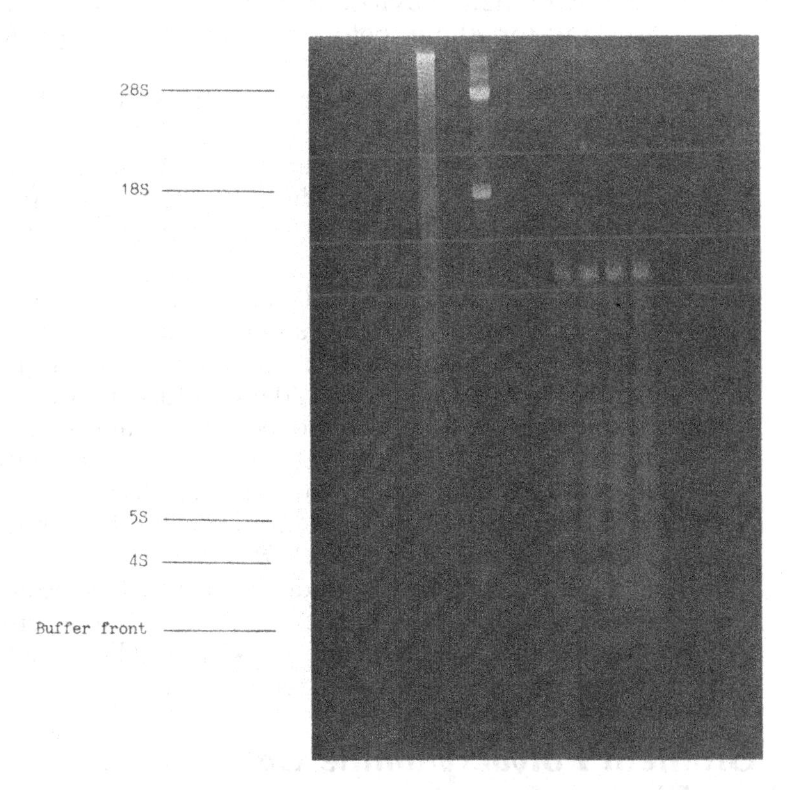


Fig. 2. A 2.4–5% (w/v) acrylamide gradient gel of various RNA samples. The gel was prepared and run as described in the Methods. The wells are numbered from left to right.

| *Well No.* | *Sample* |
|---|---|
| 5 | 10 μg poly $(A)^+$ RNA from a human cell line |
| 7 | 10 μg total cytoplasmic RNA from a human cell line |
| 10, 11, 12, 13 | Various concentrations of soluble material after a 2*M* LiCl precipitation of unbound RNA after oligo (dT)-cellulose chromatography |

## Materials

### Agarose Gel Electrophoresis

1. 100 × Electrophoresis Buffer (1*M* sodium phosphate, pH 7.0): The buffer is prepared in double distilled or distilled deionized water (dd-$H_2O$) and made 0.2% (v/v) with diethyl pyrocarbonate (DEP). This is allowed to stand for 20 min before autoclaving at 15 psi for 20 min. This treatment helps to destroy ribonuclease (RNase) activity (2) and sterilizes the solution for storage at room temperature.
2. 10 × Electrophoresis Buffer (100 m*M* sodium phosphate, pH 7.0): A tenfold dilution of 100 × electrophoresis buffer in dd-$H_2O$. Also DEP treated and autoclaved as above.
3. 2 × Sample Buffer [50% (v/v) deionized formamide, 48% (v/v) glycerol, 20 m*M* sodium phosphate, pH 7.0]: Deionized formamide is prepared by stirring 1 g of Amberlite MB-1 resin with 50 mL of formamide for 2 h. The resin beads are removed by filtration and the formamide may be stored at −70°C. After preparation a few crystals of bromophenol blue are included to act as a marker dye during electrophoresis. The buffer is stored in 1 mL aliquots at −70°C.
4. 5 mg/mL ethidium bromide: Care must be exercised when handling ethidium bromide as it is a potent carcinogen and mutagen. The stock should be stored protected from light at room temperature.

### Gradient Polyacrylamide Gel Electrophoresis

1. 15/0.75% Con Ac Bis [15% (w/v) acrylamide, 0.75% (w/v) *N,N'*-methylenebisacrylamide (bis)]: Electrophoresis grade reagents should be used, the solution filtered and stored at 4°C protected from light. Acrylamide is toxic and should be handled accordingly.
2. 30/0.8% Con Ac Bis [30% (w/v) acrylamide, 0.8% (w/v) bis].
3. 5 × Buffer E (0.18*M* Tris, 0.15*M* $NaH_2PO_4$, 5 m*M* EDTA): Store at 4°C.

4. 50% (w/v) sucrose: Autoclave and store at room temperature.
5. 10% (w/v) ammonium persulfate (AMPS): Prepared freshly for each gel.
6. 10% (v/v) *N,N,N',N'*,-tetramethylethylenediamine (TEMED): Prepared freshly for each gel.
7. Water-saturated *n*-butanol

# Method

## *Agarose Gel Electrophoresis*

1. The electrophoresis apparatus is assembled according to the manufacturers instruction. The details and quantities given below refer to a horizontal submerged gel apparatus of 11 by 14 cm giving a 3-mm thick gel. This requires 50 mL of gel mix and 800–900 mL of electrophoresis buffer.
2. One gram of agarose (BRL gel electrophoresis grade or Sigma Low EEO type), 5 mL of 10 × buffer and 45 mL of dd-$H_2O$ are mixed in a 250 mL conical flask and the agarose melted in a microwave oven or by boiling with a Bunsen burner. The mix is then allowed to cool to 60°C before pouring the gel on a level table and leaving for 1 h to set.
3. One liter of electrophoresis buffer is prepared from 10 mL of 100 × buffer and dd-$H_2O$. At this stage it is opportune to prepare the RNA samples. About 10 μg is convenient for checking the integrity of RNA but anything from 2 to 20 μg can be used. The samples (in dd-$H_2O$) are mixed with an equal volume of 2 × sample buffer and left on ice until the gel is ready.
4. The apparatus is assembled by placing the gel in the buffer tank and pouring in buffer until the gel is just submerged. The comb is removed and the wells rinsed with a syringe full of buffer. The electrophoresis system requires constant mixing of the electrolyte as a pH gradient quickly builds up because of the low buffer strength (10 m*M*). The best method is to use a pump and recirculate the buffer throughout the electrophoresis run, but if this cannot be conveniently arranged the

buffer chambers may be mixed manually every 20–30 min.

5. Electrophoresis is performed towards the anode at constant voltage for a total of 180 V-h, with a suggested maximum of 180 V. The bromophenol blue marker should have migrated about two-thirds through the gel and will be quite diffuse. The gel is stained in 1 μg/mL ethidium bromide for 30 min and may then be viewed and photographed under UV light without destaining.

## *Gradient Polyacrylamide Gel Electrophoresis*

1. The vertical slab gel apparatus is prepared for polymerization of a gel. As it is particular important that no leakage occurs during the relatively long setting time it is best to set a plug of polyacrylamide first. Quantities given below are for a 32 by 14 cm slab gel with 1.5 mm spacers.
2. The plug consists of 18% acrylamide set very rapidly with high levels of TEMED and AMPS. The gel is prepared by mixing the following solutions:

| | |
|---|---|
| 4.1 mL | 30/0.8 Con Ac Bis |
| 1.35 mL | 5 × Buffer E |
| 1.87 mL | dd-$H_2O$ |
| 100 μL | 10% AMPS |

   A suitable arrangement should be made ready for pouring the gel quickly, e.g., a 10 mL syringe with a wide-bore needle. The polymerization is started with 10 μL of undiluted TEMED and the gel poured immediately. (There is 1–2 min after adding the TEMED before the gel sets, depending on the ambient temperature.)
3. A gradient mixer is required and a suitable method for pouring the gel—gravity feed or a pump. The two gel solutions are prepared as shown in Table 1. The 5% acrylamide solution goes in the mixing chamber of the gradient maker. A check should be made to ensure that there are no trapped air bubbles between the chambers before starting pouring the gel. To start pouring the gel, first switch on the pump (or open the clip if gravity fed) and then open the tap between the two chambers. The gel is poured to within 2.5 cm of the bottom of the

Table 1
Gel Solutions for Linear Acrylamide Gradient[a]

| Stock | 5% | 1.5% | 2% Spacer |
|---|---|---|---|
| 15/0.75 Con Ac Bis | 11.7 mL | 3.5 mL | 1.33 mL |
| 5 × Buffer E | 7.0 mL | 7.0 mL | 2.0 mL |
| 50% Sucrose | 14.0 mL | — | — |
| dd-$H_2O$ | 2.2 mL | 24.3 mL | 6.6 mL |
| 10% (w/v) AMPS | 100.0 μL | 100.0 μL | 50.0 μL |
| 10% TEMED | 50.0 μL | 200.0 μL | 50.0 μL |
| Total volume | 35.0 mL | 35.0 mL | 10.0 mL |

[a]Reagents are added in the order shown above and mixed well before pouring the gel.

well-former before overlaying with about 1 mL of water-saturated *n*-butanol and leaving to set. This will take about 2 h.

4. After polymerization is complete (as shown by the appearance of a second sharp interface below the organic layer) the spacer gel is set on top. The gel mix is described in Table 1. After about 1 h of polymerization the gel is transferred to a cold room (4°C) and left for a further 30 min. Meanwhile 2 L of Buffer E is prepared and left at 4°C to cool. The samples (5–50 μg total RNA) are prepared in Buffer E/0.05% (w/v) Bromophenol Blue/10% sucrose and left on ice until the gel has completely set.
5. The apparatus is assembled for electrophoresis and the gel is pre-run at 200 V for 30 min. Power is switched off and the samples loaded with a microsyringe. Electrophoresis is at constant voltage for 5000 V-h with a suggested maximum of 400 V. The bromophenol blue should have migrated about two-thirds through the gel, but it is often difficult to judge the position of the dye because of diffusion. The gel is stained in 1 μg/mL ethidium bromide for 30 min and may then be viewed and photographed with UV light.

# Notes

1. When handling RNA every precaution should be taken to ensure that all RNase activity is destroyed or avoided. Gloves should always be worn and sterile,

DEP-treated solutions used where possible. If a syringe is used to load samples it should be well washed with ethanol and rinsed with sterile water before use. An alternative is to use sterile micropipet tips if the particular gel apparatus allows.

2. If a vertical gel apparatus is used to run the agarose gel a polyacrylamide plug should be set as described for the gradient gel except that the buffer should be 10 m*M* Na phosphate, pH 7.0. In this case the purpose of the plug is to prevent the gel slipping out of the plates. Another problem likely to be encountered with vertical agarose gels is breaking of the wells when the comb is removed. This can often be prevented by making shallow wells using a comb with teeth of about 1 cm depth.
3. Problems with RNase activity in the gels themselves do not usually occur. However, if difficulties are encountered, some measures can be taken to prevent their recurrence. With the polyacrylamide gel, 0.1% SDS may be included in the upper (cathodic) buffer chamber which reduced some RNases. The agarose gel itself and the electrophoresis buffer can be sterilized by autoclaving.
4. If trailing of the edges of the samples, and thus of the bands, is a problem 0.2% agarose should be included in the sample buffers (*3*). The reason for the improvement is unclear.

## References


1. Loening, U. E. (1967) The fractionation of high-molecular weight ribonucleic acid by polyacrylamide-gel electrophoresis. *Biochem. J.* **102,** 251–257.
2. Solymosy, F., Fedorcsak, I., Gulyas, A., Farkas, G. L., and Ehrenberg, L. (1968) A new method based on the use of diethyl pyrocarbonate as a nuclease inhibitor for the extraction of undegraded nucleic acids from plant tissue. *Eur. J. Biochem.* **5,** 520–527.
3. Shaffner, W., Gross, L., Telford, J., and Birnstiel, M. (1976) Molecular analysis of the histone gene cluster of *Psaminechinus miliaris*: II. The arrangement of the five histone-coding and spacer sequences. *Cell* **8,** 471–478.

# Chapter 13

# The Extraction of Total RNA by the Detergent and Phenol Method

***Robert J. Slater***

*Division of Biological and Environmental Sciences, The Hatfield Polytechnic, Hatfield, Hertfordshire, England*

## Introduction

Successful extraction of RNA depends on the quantitative recovery of pure nucleic acids in an undegraded form. In practice, this means that a selective extraction process is required to remove all the unwanted cellular material in a manner that minimizes degradation of the RNA by hydrolysis or ribonuclease activity. The method described here relies on cell homogenization in an aqueous medium containing a strong detergent (sodium tri-isopropylnaphthalene sulfonate) and a chelating agent (sodium 4-aminosalicylate) to solubilize the cell components. An immiscible solution of phenol is then added to selectively extract hydophobic components and to denature protein. Following phase separation, the RNA is recovered by precipitation from the aqueous phase by the addition of absolute alcohol, thereby separating the RNA from small molecular weight contaminants such as carbohydrates, amino acids, and nucleotides.

The precise conditions used in the procedure are dependent on the species of RNA required and the starting material used. For example, poly (A)-containing RNA tends to remain associated with the denatured protein during the extraction if the wrong conditions are used. The procedure described here is designed to disrupt eukaryotic nuclei and to prevent loss of poly (A)-containing RNA (*1*). It is, therefore, a useful method for the extraction of total RNA (*2*), mRNA (*1*), and the products of *in vitro* transcription (*3*). Alternative recipes are available to meet different criteria (*see* Note 2).

The detergent/phenol method is a good general procedure applicable to bacteria, fungi, and plant and animal tissues. The detergent solution described is a very effective cell lysing medium and relatively gentle homogenization procedures such as a mortar and pestle or Potter homogenizer are all that is required. Lysosyme treatment may be required however, for certain strains of bacteria (*see* Chapter 26).

The following procedure is split into two stages: the extraction of total nucleic acids followed by the removal of DNA. The latter step is optional depending on the purpose of the RNA extraction. For example, removal of DNA is a pre-requisite for in vitro translation reactions, but is not essential prior to gel electrophoresis.

Oligo (dT)-cellulose chromatography is a convenient method for the removal of DNA during the preparation of eukaryotic RNA (*see* Chapter 16); otherwise, the RNA should be incubated in a very pure solution of DNase. The method described here takes advantage of the fact that DNase I from bovine pancrease is active at 0°C and can therefore be used under conditions that inhibit RNase activity. Contaminating RNase can be removed from DNase preparations by the procedure described in Chapter 3.

## Materials

Note: Solutions containing phenol are highly toxic; gloves and safety spectacles should be worn.

1. Phenol mixture: 500 g phenol crystals
   70 mL *m*-cresol
   0.5 g 8-hydroxyquinoline
   150 mL water

   The phenol and *m*-cresol should be colorless, if not they must be redistilled. The solution is intended to be water-saturated. Store in a dark bottle at 4°C for up to 2 months. The solution darkens in color with age because of oxidation. Discard the solution if the color darkens beyond light brown. The *m*-cresol is an optional component that acts as an antifreeze and an additional deproteinizing agent.
2. Detergent solution: 1 g sodium tri-isopropylnaphthalene sulfonate (TPNS)
   6 g sodium 4-aminosalicylate
   5 mL phenol mixture

   Make to 100 mL in 50 m*M* Tris-HCl (pH 8.5). Mix the TPNS with the phenol mixture before adding the other components. Store as for phenol mixture (*see* Note 2).
3. Deproteinizing solution: phenol mixture and chloroform mixed 1:1 by volume (*see* Note 4). Store as for phenol mixture.
4. Absolute alcohol.
5. Sodium acetate buffer, 0.15*M* (pH 6.0 with acetic acid) containing 5 g $L^{-1}$ sodium dodecyl sulfate (SDS). Store at room temperature.
6. TM buffer: 50 m*M* Tris-HC1 (pH 7.4) containing 2 m*M* magnesium acetate. Autoclave and store at −20°C.
7. DNase solution: 0.5 mg $mL^{-1}$ DNase I in TM buffer. Store at −20°C in batches to avoid repeated freeze-thawing.

# Method

## *The Extraction of Total Nucleic Acids*

1. Homogenize the tissue or cells in the detergent solution (tissue:volume ratio 1:10 ideally) at 4°C or on ice.

Successful lysis of cells is accompanied by an increase in the viscosity of the solution.

2. Transfer the homogenate to a centrifuge tube (polypropylene or glass) and add an equal volume of deproteinizing solution.
3. Agitate the mixture to maintain an emulsion for 10 min at room temperature. Note that tube sealing films are not suitable for use during this process as they dissolve in the deproteinizing solution.
4. Spin the tubes for 10 min in a bench centrifuge. This separates the tube contents into three phases:

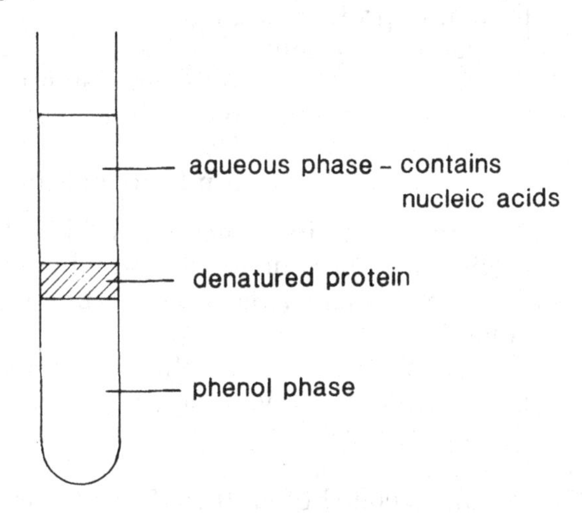


Phase separation is aided by centrifugation at 4°C, but this is not essential. If possible, spin the tubes without caps as this reduces the possibility of disturbing the contents following centrifugation.

5. Carefully, remove the upper, aqueous layer with a pipet and retain in a second centrifuge tube containing an equal volume of deproteinizing solution.
6. Re-extract the remaining phenol and protein phases by adding an additional 1 or 2 mL of detergent solution to the original centrifuge tube. Shake, centrifuge, and remove the aqueous phase as before. This step is important if a quantitative recovery of nucleic acids or total poly(A)-containing RNA is required; otherwise it may be omitted.
7. Re-extract the combined aqueous phases by shaking with the deproteinizing mixture for a further 5–10

min. Spin the tubes in a bench centrifuge and then carefully remove the aqueous phase.

8. Add 2.5 vol of absolute alcohol to the aqueous solution of nucleic acids; mix thoroughly and leave at −20°C overnight to allow precipitation of nucleic acids. A DNA precipitate resembling cotton wool often appears immediately on addition of alcohol but the RNA precipitate, resembling snowflakes, takes several hours to form at −20°C. Precipitation is more rapid at lower temperatures.
9. Collect the nucleic acid precipitate by centrifugation for 10 min in a bench centrifuge. Discard the supernatant.
10. Drain any remaining alcohol from the precipitate and then dissolve the nucleic acids in 3–4 mL of sodium acetate buffer. Add 2.5 vol of absolute alcohol, mix and precipitate the nucleic acids at −20°C overnight as before. This step is designed to remove phenol or small molecular weight contaminants present in the preparation (see note 5). If it is not convenient to spend the extra time required, it is possible to remove most of these contaminants by thoroughly washing the precipitate at room temperature, in a solution of 70% (v/v) alcohol containing 5 g $L^{-1}$ SDS.

The RNA preparation is now essentially complete, the principal contaminants being DNA, some large molecular weight carbohydrates, traces of basic proteins, and SDS. The relative importance of these contaminants varies according to the purpose of the extraction and their removal is discussed below. The purity of the nucleic acid preparation is sufficient at this stage, however, for gel electrophoresis, oligo (dT)-cellulose chromatography, or sucrose gradient fractionation (*see* Chapters 11, 12, 16, and 17, respectively).

The nucleic acids preparation can be stored as a precipitate under alcohol at −20°C or −70°C until required.

## *Removal of DNA*

11. Collect the nucleic acid precipitate by centrifugation and wash at least twice with 70% alcohol at room temperature to remove SDS.

12. Dissolve the nucleic acid preparation in 4 mL of TM buffer at 0–4°C.
13. Add 1 mL of DNase solution and incubate in an ice bath for 30 min.
14. Deproteinize with phenol/chloroform solution and precipitate the RNA with alcohol as previously described, i.e., steps 2–10 (*see* Note 6).

## Notes

1. Although phenol methods such as the procedure described here have general applications and produce nucleic acids in high yield, there are a number of points that deserve a mention. In some cases, traces of basic proteins remain associated with the nucleic acids. This can be a problem if the RNA is being used as a hybridization probe since nonspecific binding to cellulose nitrate membranes can occur. If this is a persistent problem, protein contamination can be removed by incubating the nucleic acid solution in 10 m*M* Tris-HCl (pH 7.6) containing 0.1% SDS with 100 μg/mL proteinase K for 20 min at 37°C followed by phenol extraction as previously described.

   Aggregation of RNA sometimes occurs following phenol extraction. This can be a problem during gel electrophoresis of RNA, but can be avoided by incubating the preparation in 8*M* urea at 60°C for 10 min prior to electrophoresis.
2. The conditions for phenol extraction described in this chapter are biased towards the recovery of poly(A)-containing RNA and the disruption of eukaryotic nuclei. The objectives and starting material vary greatly from one experiment to another and there are, therefore, numerous alternative recipes some of which are detailed below:

   (a) If a quantitative recovery of rRNA or tRNA is required, the detergent solution should include 60 g $L^{-1}$ sodium chloride and distilled water instead of Tris buffer.
   
   (b) If RNA is to be extracted from chloroplasts the detergent solution should contain 10 mmol $L^{-1}$ magnesium acetate.

(c) If RNA from eukaryotic nuclei is not required the detergent solution need not be so complex. Lysing solutions based on other detergents such as SDS or Nonidet-P40 are commonly used and can be substituted for the TPNS solution described here. Two such recipes are given below:

*A.* [for the extraction of mRNA from control and virus-infected mammalian tissue culture cells (*4*)]
0.15*M* NaCl
0.01*M* Tris-HCl, pH 7.9
1.5 m*M* $MgCl_2$
0.65% w/v NP40

*B.* [for the extraction of RNA from polysomes (*5*)]
0.15*M* sodium acetate
0.05*M* Tris-HCl, pH 9.0
5 m*M* EDTA
1% (w/v) SDS
20 μg/mL polyvinyl sulfate

3. In most cases the procedure described here will effectively prevent any digestion of RNA by endogenous nucleases. If it is suspected that the tissue being used is particularly rich in nucleases or the cells are lysed before addition of detergent solution, it may be necessary to use a ribonuclease inhibitor such as a vanadyl–nucleoside complex or the protein ribonuclease inhibitor from rat liver or human placenta. All of these inhibitors are commercially available and details of their use can be found in the articles by Miller et al. (*4*), Maniatis et al. (*6*), and the technical literature supplied with the product.
4. An antifoaming agent is often included in deproteinizing solutions. A commonly used recipe is to substitute chloroform with a mixture of chloroform:isoamyl alcohol, 24:1.
5. Phenol can be removed from the aqueous solution of nucleic acids, prior to alcohol precipitation, by extraction with ether. The procedure is as follows:

(a) Add an equal volume of diethyl ether (saturated with water or the extraction buffer used for the nucleic acid isolation) to the nucleic acid solution, shake, then separate the phases in a bench centrifuge.

(b) Remove the upper, ether phase containing the traces of phenol and discard.

(c) Remove traces of ether by blowing a stream of nitrogen over the surface of the solution for 10 min.

(d) Precipitate the nucleic acids with alcohol.

**Note** that diethyl ether is highly volatile and should be stored and used in a fume hood.

6. If poly(A) containing RNA is to be treated with DNase, it is wise to adjust the pH of the aqueous phase to 8.5–9.0 with Tris or NaOH prior to phenol extraction.
7. Glycogen is a common contaminant of nucleic acid preparations from mammalian tissues. Thorough washing of a nucleic acid precipitate with a solution of 3.0*M* sodium acetate (pH 7.0) will remove glycogen along with some DNA and low molecular weight RNA.

# References

1. Brawerman, G. (1974) Eukaryotic messenger RNA. *Ann. Rev. Biochem.* **43,** 621–642.
2. Slater, R. J., and Grierson, D. (1977) RNA synthesis by chromatin isolated from *Phaseoulus aureus* Roxb. The effect of endogenous nuclease. *Planta* **137,** 153–157.
3. Slater, R. J., Venis, M. A., and Grierson, D. (1978) Characterisation of RNA synthesis by nuclei isolated from *Zea mays. Planta* **144,** 89–93.
4. Miller, J. S., Roberts, B. E., and Paterson, B. M. (1982) Determination of the organisation and identity of eukaryotic genes utilising cell-free translation systems. In: *Genetic Engineering, Principles and Methods,* Vol 4, edited by Setlow, J. K., and Hollaender, A. Plenum, pp. 103–117.
5. Schleif, R. F., and Wensink, P. C. (1981) *Practical Methods in Molecular Biology.* Springer-Verlag.
6. Maniatis, T., Fritsch, E. F., and Sambrook, J. (1982). *Molecular Cloning, A Laboratory Manual.* Cold Spring Harbor, New York.

# Chapter 14

# RNA Extraction by the Proteinase K Method

***R. McGookin***

*Inveresk Research International Limited, Musselburgh, Scotland.*

## Introduction

This method of RNA extraction relies on a relatively gentle lysis procedure that should burst the cells, but leave the nuclei intact. Contamination of a relatively RNase-free cytoplasmic environment with nuclear nucleases is thus minimised. Next the polysomes are dissociated with SDS and proteinase K and finally the protein is removed by several phenol/choloroform extractions (*1,2*).

The method is simple, cheap, and rapid and is particularly useful for cells grown in culture. The details given below are based on an extraction from human neuroblastoma cells grown in standard tissue culture. This extraction procedure is not suitable for situations where there are known to be high levels of RNase present, e.g., pancreatic tissue, or where high disruptive forces have to be used so that the nuclei are liable to be lysed, e.g., most plant tissues. In these cases, the detergent and phenol method (Chapter 13) is more appropriate.

The literature contains several references where modifications to this method are suggested to improve integrity of the RNA, e.g., addition of heparin (3). If the method described here fails to give satisfactory results, then these may be tried, but it is suggested that another procedure designed for minimization of RNase, such as that using guanidine thiocyanate (*see* Chapter 15), is used rather than spending time and resources unnecessarily.

## Materials

Analar grade reagents and double distilled or distilled deionized (dd-$H_2O$) should be used for all solutions. All buffers are treated with 0.2% (v/v) diethyl pyrocarbonate for 20 min and autoclaved. Buffers containing Tris should be checked for changes in pH after this treatment.

1. ISO-TKM 150 m*M* KCl, 10 m*M* Tris-Cl, pH 7.5., 1.5 m*M* $MgCl_2$.
2. HYPO-TKM 10 m*M* KCl, 10 m*M* Tris-Cl, pH 7.5., 1.5 m*M* $MgCl_2$.
3. 10% (v/v/) Nonidet P-40.
4. 10% (w/v) SDS autoclaved at 10 psi for 10 min only
5. 25 mg/mL Proteinase K. The enzyme is dissolved in dd-$H_2O$ and self-digested at 37°C for 1 h before storing at −20°C.
6. 3*M* KCl.
7. TKE 300 m*M* KCl, 10 m*M* Tris-Cl, pH 7.5, 1 m*M* EDTA.
8. Redistilled phenol saturated with TKE: Distillation of phenol is a potentially dangerous operation. It is best to consult an experienced person if you are unfamiliar with this technique. The phenol is made 0.1% (w/v) 8-hydroxyquinoline, saturated with 30 mL TKE/100 mL phenol and stored at −20°C.

## Method

1. The cells are harvested as quickly as possible onto ice. They are pelleted at 1000*g* for 10 min at 4°C and washed several times in ice-cold ISO-TKM. (The exact number

of times will depend on the amount of material harvested, but should be at least three times.)

2. The cells are finally resuspended in cold HYPO-TKM at about $10^8$ cells/mL and transferred to a polypropylene centrifuge tube. The suspension is made 0.5% (v/v) Nonidet. The debris is pelleted at 15,000*g* for 10 min at 4°C.
3. The supernatant is removed into a fresh tube and made to 0.5% SDS and 200 μg/mL proteinase K. After incubation at 37°C for 30 min, 1/10 vol of 3*M* KCl and ½ vol each of phenol and chloroform are added. The tube is capped with some organic-resistant material and whirlimixed vigorously. A 500*g* spin for 10 min at 4°C is used to break the emulsion and separate the phases.
4. The upper aqueous phase is removed, taking care not to disturb the denatured protein at the interface, into a fresh tube on ice and the organic layer is re-extracted with 1 vol of TKE. The aqueous phase may be cloudy because of SDS, but this is not important. A more important problem may be incomplete separation of the phases. Warming to 30°C may help to break these up.
5. The pooled aqueous phase is extracted twice more with 0.5 vol each of phenol and choloroform and finally once with chloroform alone. The final aqueous phase is precipitated with 2.5 vol of ethanol at −20°C overnight.
6. Next day the RNA is spun down at 12,000*g* for 20 min. The white pellet is washed once with 80% ethanol, dried *in vacuo* and redissolved in dd-$H_2O$. Normally the yield of RNA is measured by a scan of the spectrum from 220 to 320 nm (1 OD Unit at 260 nm is equivalent to 45 μg/mL RNA). This also indicates any protein contamination problems.

## Notes

1. When handling RNA every effort must be made to ensure that no RNase contamination occurs. The solutions should be DEP-treated and autoclaved, as described in the materials section. Where possible, all glassware should be oven-baked and plasticware autoclaved at 15 psi for 20 min. Gloves must be worn both

to stop contamination from skin RNases and to protect the personnel from the caustic reagents.

2. In the case of the neuroblastoma cells described above, no mechanical homogenization of the cells is necessary. In circumstances where the cells prove more difficult to lyse, a few strokes with a glass homogenizer should be sufficient. The lysis can be checked with Trypan Blue staining.
3. Once the extraction has been started, i.e., the cells harvested, the whole process should be carried out as quickly as possible up to the first phenol extraction. The cells must not be allowed to sit on ice too long, nor should the washes with ISO-TKM be carried out at elevated temperatures. However, as mentioned in the introduction, if this RNA extraction method repeatedly gives low yields and/or shows degradation of the products, an alternative procedure should be tried.

## References

1. Wiegers, U. and Hilz, H. (1971) A new method using 'proteinase K' to prevent RNA degradation during isolation from HeLa cells. *Biochem. Biophys. Res. Commun.* **44,** 513–519.
2. Perry, R. P., LaTorre, J., Kelley, D. E., and Greenberg, J. R. (1972) On the lability of poly(A) sequences during extraction of messenger RNA from polyribosomes. *Biochim. Biophys. Acta* **262,** 220–226.
3. Morrison, M. R., Baskin, F., and Rosenberg, R. N. (1977) Quantitation and characterisation of poly(A)-containing messenger RNAs from mouse neuroblastoma cells. *Biochim. Biophys. Acta* **476,** 228–237.

# Chapter 15

# RNA Extraction by the Guanidine Thiocyanate Procedure

*R. McGookin*

*Inveresk Research International Limited, Musselburgh, Scotland*

## Introduction

This method relies on the strong chaotropic nature of the reagents involved to completely denature any ribonuclease (RNase) present in the sample. After lysis in guanidine thiocyanate buffers there are two possibilities for isolation of the RNA. One method involves a series of differential precipitation steps in guanidine hydrochloride (*1*). The alternative, detailed here, involves centrifugation of the samples on a cushion of 5.7*M* CsCl (*2,3*). The RNA passes through this cushion, whereas the DNA and the majority of other cellular macromolecules remain above the cushion.

The description below is based on an extraction from 3-d germinated cucumber seeds, a tissue undergoing much metabolic reorganization and containing large amounts of nuclease activity (*4*). The tissue is also tough so fairly strong disruptive procedures are needed. Other tissues, such as cells in culture, simply require to be resuspended in the guanidine thiocyanate buffer to lyse the cells and release the RNA.

## Materials

1. Thiocyanate buffer: (5*M* guanidine thiocyanate, 50 m*M* Tris-Cl, pH 7.5, 10 m*M* EDTA, 5% 2-mercaptoethanol). This buffer is prepared freshly each time from solid thiocyanate, sterile stocks of 1*M* Tris-Cl, pH 7.5, and 0.5*M* EDTA, pH 7.5, and undiluted mercaptoethanol. It is filtered through a 22 μm filter into sterilized glassware.
2. CsCl cushion (5.7*M* CsCl, 100 m*M* EDTA, pH 7.5). The solution is filtered through a 22 μm filter and stored at 4°C.
3. 10 m*M* Tris Cl, pH 7.5.
4. 6*M* Ammonium acetate.

## Method

1. The tissue is harvested into liquid nitrogen in a precooled mortar and ground to a fine powder. This material is mixed with about 4 vol of thiocyanate buffer (if the material is very viscous, more buffer is added) then poured into a tight-fitting homogenizer and given several strokes until a smooth, creamy textured solution results.
2. Solid *N*-lauroyl sarcosine is added to give a final concentration of 4% (w/v) and CsCl to 0.15 g/mL. After these have dissolved large debris is removed by centrifugation at 15,000*g* at 4°C for 20 min.
3. While the homogenate is spinning, the required number of ultracentrifuge tubes are made ready with CsCl cushions in the bottom. In the case of Beckman SW 50.1 tubes, each will hold 4 mL of homogenate and 1 mL of 5.7*M* CsCl cushion. After the low-speed spin the supernatant is carefully layered over the cushions and the tubes loaded into the rotor. The RNA is pelleted at 100,000*g* (32,500 rpm in SW 50.1 rotor) for 18 h at 20°C.

4. The tubes are carefully removed from the buckets and the homogenate is aspirated off through a Pasteur pipet. At the interface will be a layer of very "stringy" material, which is the DNA. The walls of the tube and the surface of the cushion are carefully washed three times with sterile water and the tube inverted to drain off the cushion. The pellet should have a clear, lens-like appearance although there are sometimes "frilly edges" caused by carbohydrate material.
5. The pellet is carefully dissolved in 10 m*M* Tris-Cl, pH 7.5. This can be a slow process since the high salt concentration makes the RNA reluctant to dissolve. The process can be speeded up by repeatedly sucking and ejecting the solution with a micropipet and sterile tip. The solution is made 4% (v/v) with 6*M* ammonium acetate, 2 vol of ethanol added and left at −20°C overnight to precipitate the RNA.

## Notes

1. The thiocyanate buffer is particularly toxic and great care must be exercised when handling it. The material should be handled in the fume cupboard whenever possible.
2. The same criteria of cleanliness and sterility apply to this method as applied in Chapter 14. Theoretically the solutions and glassware need not be sterile until after the RNA is separated from the thiocyanate buffer. However, it is suggested that one use sterile stocks and glassware at all times.
3. A critical stage in this extraction method is the removal of the homogenate and washing of the cushion surface after ultra-centrifugation. Although the RNases are completely denatured in the thiocyanate buffer once this is diluted out, they can renature and regain their activity. Some authors suggest washing the tube walls with dilute diethyl pyrocarbonate, but since this reagent can degrade RNA (*5*), this is not recommended. An alternative is to cut the top portion off the centrifuge tube.

## References

1. Chirgwin, J. M., Przybyla, A. E., Macdonald, R. J., and Rutter, W. J. (1979) Isolation of biologically active ribonucleic acid from sources enriched in ribonuclease. *Biochemistry* **18,** 5294–5299.
2. Glisin, V., Crkvenjakov, R., and Byus, C. (1974) Ribonucleic acid isolation by cesium chloride centrifugation. *Biochemistry* **13,** 2633–2637.
3. Kaplan, B. B., Bernstein, S. L., and Gioio, A. E. (1979) An improved method for the rapid isolation of brain ribonucleic acid. *Biochem J.* **183,** 181–184.
4. Becker, W. M., Leaver, C. J., Weir, E. M., and Riezman, H. (1978) Regulation of glyoxysomal enzymes during germination of cucumber. I. Developmental changes in cotyledonary protein, RNA and enzyme activities during germination. *Plant Physiol.* **62,** 542–549.
5. Wiegers, U., and Hilz, H. (1971) A new method using 'proteinase K' to prevent mRNA degradation during isolation from HeLa cells. *Biochem. Biophys. Res. Commun.* **44,** 513–519.

# Chapter 16

# The Purification of Poly(A)-Containing RNA by Affinity Chromatography

*Robert J. Slater*

*Division of Biological and Environmental Sciences, The Hatfield Polytechnic, Hatfield, Hertfordshire, England*

## Introduction

The vast majority of eukaryotic mRNA molecules contain tracts of poly(adenylic) acid, up to 250 bases in length, at the 3′ end. This property is very useful from the point of view of mRNA extraction because it forms the basis of a convenient and simple affinity chromatography procedure (*1*). Under high salt conditions (0.3–0.5*M* NaCl or KCl), poly(A) will hybridize to oligo(dT)-cellulose or poly(U)-Sepharose. These commercially available materials consist of polymers of about 10–20 nucleotides, covalently bound to a carbohydrate support, and bind RNA containing a poly(A) tract as short as 20 residues. Ribosomal and transfer RNAs do not possess poly(A) sequences and will not bind (*see* Note 1).

Following thorough washing of the column, mRNA can be recovered by simply eluting with a low salt buffer. This is more difficult in the case of poly(U)-Sepharose columns because poly(A) binds more tightly to this ligand and stronger elution conditions, such as the inclusion of formamide in the buffers, are required. For this reason oligo(dT)-cellulose has been chosen here.

The procedure includes an optional heat-treatment step to reduce aggregation of RNA prior to chromatography (2) and SDS is present in the buffers as a precaution against ribonuclease activity. The detergent must be removed, however, before in vitro translation of the mRNA (*see* Chapters 19–21).

## Materials

1. Oligo (dT) binding buffer: 20 m*M* Tris-HCl (pH 7.5), 1 m*M* EDTA, 0.5*M* NaCl, 0.2%(w/v) SDS.
2. Oligo (dT) binding buffer (×2 salt concentration): 20 m*M* Tris-HCl (pH 7.5) 1 m*M* EDTA, 1.0*M* NaCl, 0.2%(w/v) SDS.
3. Oligo (dT) elution buffer: 20 m*M* Tris-HCl (pH 7.5), 1 m*M* EDTA, 0.2% (w/v) SDS.
4. 2.0*M* Sodium acetate (pH 6.0 with acetic acid).
5. 0.3*M* NaOH.
6. A small column packed with 25 mg to 1 g of oligo (dT) cellulose [1 g will bind 20–40 $A_{260}$ units of poly(A)] swollen in elution buffer and autoclaved at 115°C for 20 min. A conveniently small column can be made from a Pasteur pipet or an automatic pipet tip plugged with siliconized glass wool.

   All solutions should be autoclaved and all glassware heat-treated at 200°C for an hour before use to avoid contamination with ribonuclease.

## Method

Unless otherwise stated, all operations are carried out at room temperature.

1. Thoroughly drain ethanol from the nucleic acid precipitate and dissolve the pellet in oligo (dT) elution buffer.

An RNA concentration of 200 µg/mL is convenient but not critical.

2. Heat the solution for 5 min at 65°C, cool, and add an equal volume of oligo (dT) binding buffer (×2 salt concentration).
3. Equilibrate the column with oligo (dT) binding buffer and load the sample at a rate of approximately 10 mL/h. The column effluent can be reapplied to the column, if desired, to ensure maximum binding of poly(A)-containing RNA.
4. Wash the column with oligo (dT) binding buffer (at a faster rate than during loading if desired) until the $A_{260}$ reading of the effluent is at a minimum.
5. Elute the column with oligo (dT) elution buffer to recover poly(A)-containing RNA and pool all the UV absorbing fractions. If desired, the RNA can be further purified by rechromatography on a clean oligo-(dT) column by returning to step 2.
6. Adjust the salt concentration of the eluate to 0.15*M* sodium acetate by addition from the stock (2*M*) solution, add 2.5 vol of absolute ethanol and precipitate the nucleic acids overnight at −20°C.
7. Collect the RNA precipitate by centrifugation and wash at least twice with 70% alcohol at room temperature to remove SDS. Store as a precipitate under alcohol at −20°C (or lower) or drain and dissolve in sterile distilled water for in vitro translation.
8. After use, the column can be cleaned by eluting with 0.3*M* NaOH and then oligo-(dT) elution buffer containing 0.02% (w/v) sodium azide until the pH returns to 7.5. The column can be stored at room temperature until required again. The column can be re-used indefinitely.

# Notes

1. In practice, some rRNA binds to the column at 0.5*M* salt. This can be removed, prior to mRNA elution, by washing the column with buffer containing 0.1*M* NaCl. Alternatively, the poly(A)-containing RNA can be rechromatographed as suggested in step 5. Small quantities of rRNA do not, however, significantly affect in vitro translation of mRNA.

## References

1. Aviv, H. and Leder, P. (1972) Purification of biologically active globin messenger RNA by chromatography on oligothymidylic acid cellulose. *Proc. Nat. Acad. Sci. USA* **69,** 1408–1412.
2. Nakazato, M., and Edmonds, M. (1974) Purification of messenger RNA containing poly(A) sequences. *Methods Enzymol.* **29,** 431–443.

# Chapter 17

# Messenger RNA Fractionation on Neutral Sucrose Gradients

*R. McGookin*

*Inveresk Research International Limited, Musselburgh, Scotland*

## Introduction

The separation of RNA on the basis of size by sucrose gradient fractionation is a technique frequently employed in a cloning strategy (*1–3*). After production of poly(A)$^{+}$ mRNA by affinity chromatography (*see* Chapter 16) the RNA can be fractionated once or twice on sucrose gradients to produce a subpopulation of mRNA enriched for a particular species. The RNA fractions from the gradient can be assayed by in vitro translation and subsequent immunoprecipitation or bioassay of the products (*see* Chapters 19–22).

The method described below is based upon separation of RNA on a 5–20% (w/v) sucrose gradient in a nondissociating buffer. This method is simpler to use than including formamide in the gradients (4). The removal of the formamide requires multiple precipitation steps with associated loss of material. Unfortunately, in some cases, the mRNA species of interest may appear in a very broad range of fractions when nondissociating gradients are used. In such cases formamide containing gradients are unavoidable.

## Materials

1. 10 × Gradient Buffer: (100 m*M* Tris-Cl, pH 7.5; 1 *M* KCl; 10 m*M* EDTA). The buffer is prepared from distilled deionized water (dd-$H_20$) and treated with 0.2% (v/v) diethylpyrocarbonate (DEP) before autoclaving. The pH should be checked as DEP reacts with Tris base.
2. Light sucrose [5% (w/v) sucrose in 1× gradient buffer]. This solution should be autoclaved for no more than 15 min at 15 psi maximum as otherwise the sucrose will caramelize.
3. Heavy sucrose [25% (w/v/) sucrose in 1× gradient buffer]. Autoclave as for light sucrose.

## Method

1. As with all methods involving RNA, the components that come into contact with the RNA should be sterile and RNase-free if possible. The centrifuge tubes and silicon tubing used to pour the gradients should be autoclaved. The gradient former should preferably be made of glass and oven-baked before use. When the only available gradient mixer is plastic, it should be soaked in a 0.5% (v/v) DEP solution and thoroughly rinsed with sterile dd-$H_20$ before use.
2. The volumes described here relate to 14 mL tubes for a MSE 6 × 14 mL Titanium Swing Out rotor. A template tube is filled with 12 mL of water, a piece of silicon tubing placed in it, and set alongside the tubes in which the gradients are to be formed. 8 mL of light sucrose is placed in the mixing chamber of the gradient maker

and 8 mL of heavy sucrose in the other chamber. The connecting valve is opened briefly to dislodge trapped air bubbles. The outflow tube from the gradient mixer is placed in the bottom of a centrifuge tube and the gradient poured. When the liquid level has reached that in the template tube the flow of sucrose is stopped and the silicon tubing carefully removed from the centrifuge tube. This process is repeated until the required number of tubes have been prepared and the gradients then stored at 4°C.

3. The samples are prepared in 10 m*M* Tris-Cl, pH 7.5, 100 m*M* KCl, 1 m*M* EDTA (gradient buffer). About 200 μg of total RNA can be used as a marker. For separation of poly(A)$^+$ RNA a useful starting value is 100 μg. The samples are heated to 65°C for 2 min and snap cooled on ice for 5 min before carefully laying them on top of the gradients.
4. The tubes are balanced and loaded into the rotor. Centrifugation is carried out at 4°C for 4 h at 40,000 rpm (200,000*g*). It is best, however, to start the run slowly until the rotor and centrifuge temperatures have equilibrated. Thus a 10 min first stage at 5000 rpm is used. Deceleration may be with the brake on without significantly disturbing the separation.
5. After centrifugation the marker gradient(s) may be fractionated on a commercially available apparatus with an integral UV monitor. The sucrose concentration of the fractions is measured with a refractometer. The mRNA gradients, or other samples where recovery is important, are best fractionated by gently removing 250 μL fractions from the tube with a micropipet and sterile tips. This avoids possible exposure to RNases within the fractionation apparatus. When 100 μg of material have been loaded the RNA samples can be precipitated directly with 2.5 vol of ethanol at −20°C.

## Notes

1. This is a fairly straightforward technique with few associated problems. The most likely cause of failure is the introduction of RNase at some stage. However, if the

precautions mentioned in the Method section are adhered to, degradation should seldom, if ever, occur.

2. Using different types of rotor and centrifuge will involve some modification to the speed and duration of the run. One or two preliminary trials with markers on the gradient should be sufficient to determine optimum conditions. In general, long narrow tubes give better separation than short, wide ones.

## References

1. Marcu, K. B., Valbuena, O., and Perry, R. P. (1978) Isolation, purification and properties of mouse heavy-chain immunoglobulin mRNAs. *Biochemistry* **17,** 1723–1733.
2. Vamvakopoulos, N. C., and Kourides, I. A. (1979) Identification of separate RNA's coding for the alpha and beta subunits of thyrotropin. *Proc. Natl. Acad. Sci. USA* **76,** 3809–3813.
3. Katcoff, D., Nudel, U., Zevin-Sonkin, D., Carmon, Y., Shani, M., Lehrach, H., Frischauf, A. M., and Yaffe, D. (1980) Construction of recombinant plasmids containing rat muscle actin and myosin light chain DNA sequences. *Proc. Natl. Acad. Sci. USA* **77,** 960–964.
4. Brown, D. B., and Suzuki, Y. (1974) The purification of the messenger RNA for silk fibroin. *Meth. Enzymol.* **30,** 648–654. Eds. Moldave, K., and Grossman, L., Academic Press, New York.

# Chapter 18

# The Estimation of mRNA Content by Poly(U) Hybridization

***Robert J. Slater***

*Division of Biological and Environmental Sciences, The Hatfield Polytechnic, Hatfield, Hertfordshire, England*

## Introduction

The sequence of poly(adenylic) acid, present at the 3′ end of the majority of eukaryotic mRNA molecules, forms the basis of a sensitive technique for the estimation of mRNA content in nucleic acid samples. Under suitable conditions, poly(A) will form RNA–RNA hybrids with poly(U) in vitro. The poly(A) content of RNA samples can therefore be detected by hybridization with saturating amounts of $^3$H-poly(U) (*1,2*). Following the removal of excess $^3$H-poly(U) by ribonuclease treatment, the hybrids can be collected by TCA precipitation and quantified by scintillation counting. If the results are compared with data obtained from a parallel experiment using known amounts of poly(A), a value for the poly(A) content of any number of RNA preparations can be obtained. The technique can be used to detect less than 10 $^{10}$g of poly(A).

To obtain accurate and reliable results it is important to confirm that saturation of the hybridization reaction has occurred. Ideally, a saturation experiment should be carried out, using increasing concentrations of $^{3}$H-poly(U), for every experimental sample to be investigated. In practice, this may be difficult and expensive if many unknown samples are being tested. The procedure presented here, therefore, consists of two experiments to characterize the hybridization reaction with standard poly(A) and an experimental RNA sample, and a third experiment to determine the poly(A) content of any number of RNA samples. All three experiments can, however, be carried out simultaneously. The procedure is very simple and no specialist equipment is required. Some examples of the kind of data that can be obtained are given in Figs. 1–3.

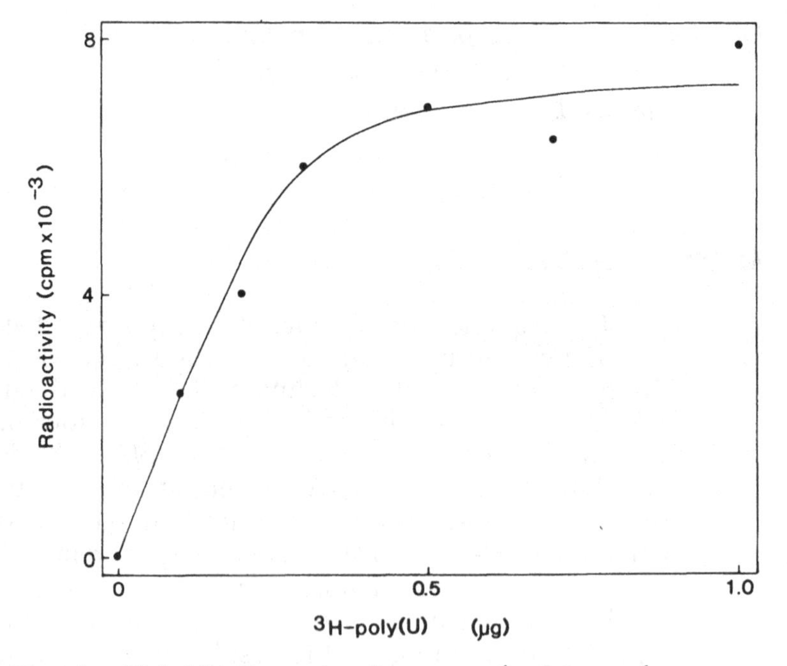


Fig. 1. Hybridization saturation curve for 0.1 μg of commercial poly(A) with increasing amounts of $^{3}$H-poly(U).

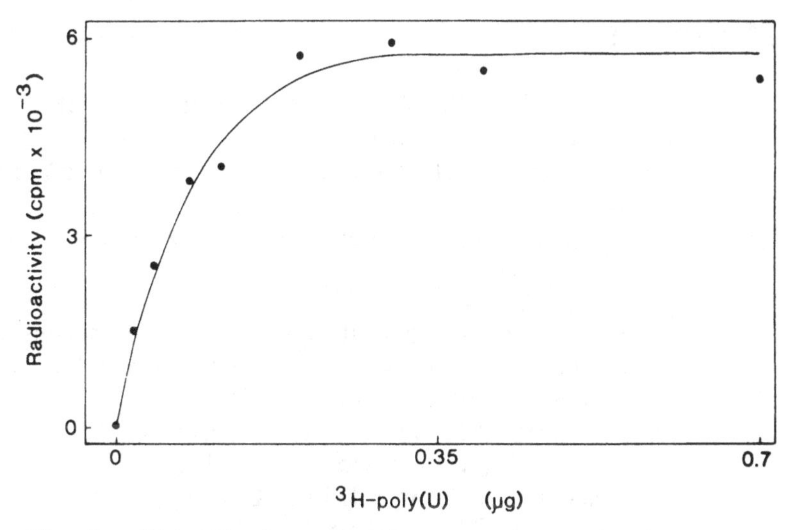


Fig. 2. Hybridization saturation curve for 50 µg of total RNA (prepared from embryo axes of *Acer pseudoplatanoides* seeds according to the procedure described in Chapter 13) with increasing amounts of $^{3}$H-poly(U).

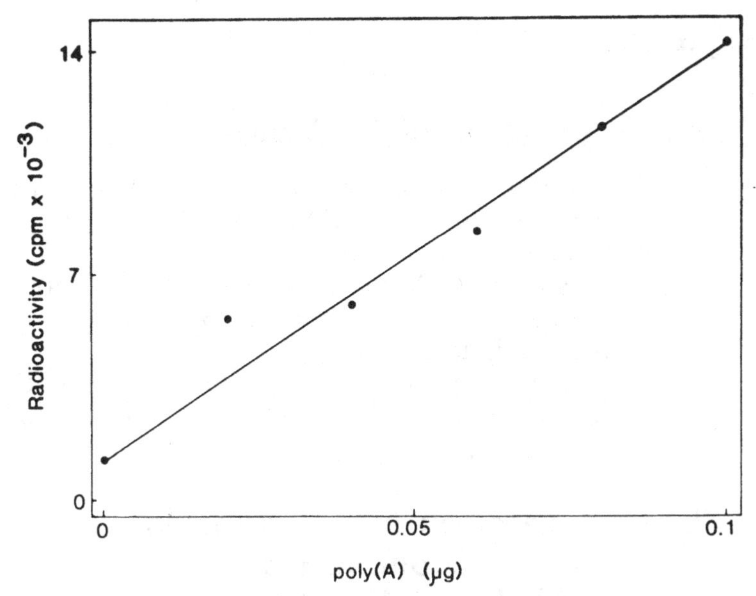


Fig. 3. Calibration plot following hybridization of excess $^{3}$H-poly(U) to increasing amounts of poly(A).

## Materials

1. SSC (×2): 0.3*M* NaCl, 0.03*M* Na citrate, to pH 7.0 with citric acid and autoclaved.
2. Standard poly(A) solution: 10 μg/mL poly(A) in SSC ×2.
   Store in a polypropylene or siliconized glass container at −20°C.
3. $^3$H-poly(U) (500 mCi/mmol): 10 μg/mL in SSC ×2. Higher specific activity can be used, if necessary, to increase sensitivity. Store in a polypropylene or siliconized glass container at −20°C.
4. Pancreatic RNase A, 1 mg/mL in SSC ×2.
5. 2′ and 3′ uridylic acid, 1 mg/mL in SSC ×2.
6. Yeast RNA, 1 mg/mL in SSC ×2.
7. 10% (w/v) TCA, ice-cold.
8. 5% (w/v) TCA containing 0.1% (w/v) 2′ and 3′ uridylic acid, ice-cold.
9. Absolute ethanol.
10. Glass fiber filters, 2.5 cm (e.g., Whatman GF/C) and filter tower apparatus.

## Method

### The Preparation of RNA Samples

1. The RNA samples must be free of DNA. If DNA is present, treat with DNase solution as described in Chapter 13.
2. Wash the RNA precipitate with 70% alcohol at room temperature to remove any SDS. Centrifuge and drain the precipitate of alcohol.
3. Dissolve the RNA in SSC ×2 and adjust the RNA concentration to 1 mg/mL. ($A_{260}$ of 1 in a 1-cm light path ≃ 45 μg/mL).

### The Construction of a Saturation Curve with Commercial Poly(A) (Procedure 1)

4. Set up six tubes, in duplicate, containing 0.1 μg poly(A).

5. Add to the six pairs of tubes, 0.1, 0.2., 0.3, 0.5, 0.7, and 1.0 µg $^{3}$H-poly(U), respectively, and make up to 2 mL with SSC ×2.
6. Swirl the tubes, then incubate for 3 min at 90°C, followed by 60 min at 25°C.
7. Add 60 µL of the RNase solution, mix, and incubate for 20 min at 25°C.
8. Add 100 µg yeast RNA, 100 µg 2′ and 3′ uridylic acid and 4 mL 10% TCA. Leave on ice for 2 h.
9. Collect the precipitate on glass fiber discs, pre-wetted with 5% TCA, under vacuum. Wash the discs with 3 × 10 mL of the 5% TCA solution followed by 2 × 5 mL absolute ethanol.
10. Dry the filters and estimate the radioactivity in a scintillation counter.

### *The Construction of a Saturation Curve with Experimental RNA (Procedure 2)*

11. Proceed through steps 4–10, as for the saturation curve with commercial poly(A), substituting the 0.1 µg poly(A) in step 4 with 50 µg of the experimental RNA (see note 1).

### *The Estimation of Poly(A) Content of Experimental Samples (Procedure 3)*

12. Set up six tubes, in duplicate, containing 0–0.1 µg standard poly(A) and additional tubes containing 50 µg or less of the experimental RNA samples. Any number of experimental samples can be analyzed, but they should be tested at least in duplicate.
13. Add 0.3 µg $^{3}$H-poly(U) to all the tubes and adjust the total volume to 2 mL with SSC × 2 (*see* Note 2).
14. Proceed through steps 6–10.

## Treatment of Results

To calculate the results, plot a graph of radioactivity collected on the filters against $^{3}$H-poly(U) concentration, for the experiments containing 0.1 µg poly(A) and 50 μg RNA. Examples of the kind of data that can be expected are shown in Figs. 1 and 2.

If saturation is not obtained, check the following:

(i) That the amounts of poly(A) and $^3$H-poly(U) added were correct.
(ii) That the RNA sample was free of DNA and protein.
(iii) That the excess poly(U) was degraded by RNase.
(iv) That the filters were sufficiently washed.

If satisfactory saturation is obtained, use the data from procedure 3 to plot a calibration curve of radioactivity bound to the filters against standard poly(A) content (Fig. 3). Values for the poly(A) content of unknown samples can then be obtained by reference to this graph.

## Notes

1. The amount of experimental RNA used is not critical providing the poly(A) content is low enough to ensure saturation of the hybridization reaction by $^3$H-poly(U). The figure of 50 μg suggested is based on the assumption that the sample tested is total, cellular RNA. If a very low poly(A) content of the sample is expected, carry out additional hybridization reactions containing 0.02–0.1 μg $^3$H-poly(U) to ensure that full details of the saturation curve can be recorded.
2. The precise amount of $^3$H-poly(U) is chosen from the data obtained from procedures 1 and 2 and should be at least three times the theoretical amount required to cause saturation of the hybridization reaction. The figure of 0.3 μg suggested here should be adequate for most experiments, but in many cases it may be possible to use less.

## References

1. Covey, S. N., and Grierson, D. (1976) The measurement of plant polyadenylic acid by hybridization with radioactive polyuridylic acid. *Planta* **131,** 75–79.
2. Bishop, J. O., Rosbash, M., and Evans, D. (1974) Polynucleotide sequences in eukaryotic DNA and RNA that form ribonuclease-resistant complexes with polyuridylic acid. *J. Mol. Biol.* **85,** 75–86.

# Chapter 19

# DNA Directed In Vitro Protein Synthesis with *Escherichia coli* S-30 Extracts

***Jytte Josephsen and Wim Gaastra***

*Department of Microbiology, The Technical University of Denmark, Lyngby, Denmark*

## Introduction

A DNA-directed cell-free protein synthesizing system was originally developed by Zubay (*1*). The system contains a crude extract prepared from *Escherichia coli*. This extract contains the machinery necessary for the transcription and translation, i.e., ribosomes and RNA polymerase. To this system, it is necessary to add all 20 amino acids, all four ribonucleotide triphosphates, transfer RNA, an energy generating system, and various salts. The DNA template is incubated with this mixture for at least 30 min at 37°C before gene products are examined. The following method is essentially as described by Zubay (*1*), but with minor modifications as described by Valentin-Hansen et al. (2).

# Materials

1. Growth medium:

| | | |
|---|---|---|
| Yeast extract | 10 | g |
| $KH_2PO_4$ | 5.6 | g |
| $K_2HPO_4 \cdot 3H_2O$ | 37.8 | g |
| $MgSO_4 \cdot 7H_2O$ | 492 | mg |
| Thiamin | 10 | mg |
| Glucose | 10 | g |

   To 1 L with distilled water. The glucose is added after autoclaving.
2. Buffer A: 0.01M Tris acetate, pH 8; 0.014*M* magnesium acetate; 0.06*M* potassium acetate; 0.001*M* dithiothreitol.
3. *E. coli* strain containing a chromosomal deletion of the gene being examined.
4. In vitro mixture (*see* Table 1).

# Methods

## *Preparation of the S-30 Extract*

1. Grow the bacteria at 30°C in 5 L of the growth medium. Inoculate the medium with cells grown overnight on the same medium. Start the culture at an optical density at 436 nm of 0.2. It is important that the cells get a lot of oxygen during growth to ensure maximal growth.
2. Stop the cell growth at an $OD_{436}$ of 7–8 by a quick cooling of the culture. Do this by placing the vessel with the culture in an ice bath and add at the same time 2–3 L of ice directly into the culture. It is very important that the cells get a lot of air under this procedure.
3. Harvest the cells in a continuous flow centrifuge which is cooled to −10°C.

4. After harvest resuspend the cells in 100 mL of buffer A and centrifuge for 20 min at 7000*g*. Rewash the cells using the same procedure as before. Weigh the cells and store them in liquid $N_2$ until required.
5. To prepare an S-30 extract, the frozen cells are allowed to thaw at room temperature for 20 min. Suspend the cells in 1.4 mL buffer A/g cells.
6. Perform the following procedures at 0–4°C unless otherwise noted. Lyse the cell suspension in a French pressure cell using pressure of 6000–8000 psi. Add 100 μL, 0.1*M* dithiothreitol per 10 mL of extract.
7. Centrifuge the extract for 30 min at 30,000*g* at 4°C. Measure the volume of the supernatant.
8. In order to remove messenger RNA, which may be present, incubate the supernatant for 80 min at 37°C with the following mixture: Add to 10 mL of supernatant:

| | |
|---|---|
| 1.0*M* Tris acetate, pH 8.0 | 1.000 mL |
| 1.4*M* magnesium acetate | 0.021 mL |
| 0.22*M* ATP, pH 7.5 | 0.036 mL |
| 0.42*M* phosphoenolpyruvate, pH 7.9 | 0.215 mL |
| 0.1*M* dithiothreitol | 0.130 mL |
| Mixture of the 20 amino acids, each 2.5 m*M* | 0.002 mL |
| Pyruvate kinase, 10 mg/mL | 0.010 mL |
| $H_2O$ | 1.580 mL |

9. Dialyze for 4 h against 1 L of buffer A. Change the buffer 4 times.
10. Freeze the ready extract in small portions (300–500 μL) and store them at −80°C or in liquid nitrogen. Thaw the extract at 4°C just before use. The extracts remain active for at least 1 yr in liquid nitrogen.

## *In Vitro Synthesis*

1. Prepare a freshly mixed in vitro mixture with the composition shown in Table 1.
2. Distribute to different tubes DNA at appropriate concentration (~ 5–10 μg/mL, final concentration) Add distilled water to a final volume of 45 μL.

Table 1
Composition of the In Vitro Mixture

| Component | | μL | Concentration in reaction mix |
|---|---|---|---|
| Tris acetate | 2.2 *M*, pH 8.0 | 100 | 44 m*M* |
| K-acetate | 2.75 *M* | 100 | 55 m*M* |
| $Ca(Ac_2)$ | 0.37 *M* | 100 | 7.4 m*M* |
| $NH_4Ac$ | 1.35 *M* | 100 | 27 m*M* |
| $Mg(Ac_2)$ | 0.50 *M* | 50 | 5 m*M* |
| Folinic acid | 3 mg/mL | 50 | 30 μg/mL |
| tRNA | 10 mg/mL | 50 | 0.1 mg/mL |
| ATP | 0.22 *M*, pH 7.5 | 50 | 2.2 m*M* |
| CTP | 0.11 *M*, pH 7.5. | 25 | 0.55 m*M* |
| UTP | 0.11 *M*, pH 7.5. | 25 | 0.55 m*M* |
| GTP | 0.11 *M*, pH 7.5. | 25 | 0.55 m*M* |
| 20 Amino acids | 0.05 *M* | 65 | 0.25 m*M* |
| DTT | 0.7 *M* | 10 | 1.4 m*M* |
| PEP | 0.42 *M*, pH 7.5 | 250 | 21 m*M* |
| PEG | 40% | 250 | |
| | | 1250 μL | |

3. Add 30 μL of S-30 extract to each tube.
4. Prewarm the above test tubes and the in vitro mix for 3 min at 37°C while shaking rapidly.
5. Start the in vitro synthesis by adding 25 μL in vitro mixture to each tube. Mix well. While shaking, the incubations are continued for at least 30 min at 37°C. Stop the reaction by placing the tubes in an ice bath.
6. The solutions may now be assayed for the presence of gene products (see notes section for a suitable test system).

# Notes

1. The described cell-free protein synthezising system is very useful for studying the regulation of protein synthesis at the level of transcription and translation. It is easy to determine which low molecular effectors, e.g., cyclic AMP and inducers, have an influence on the protein synthesis.

A necessary way of testing ones system for regulating proteins is to prepare the S-30 extract from a wild-type and from a strain with a mutation in the gene encoding the regulating protein—a repressor or an activator—but the best way to study regulation is to add to the system a purified regulating protein.

2. To test whether the system is working use DNA containing the *lac* operon and measure the β-galactosidase activity (*3*). The system is working well if it develops a yellow color within 10 min.
3. The optimum DNA concentration should be determined prior to definitive experiments by varying the DNA concentration while keeping other variables constant. The amount of DNA to be added is dependent on the sensitivity of the method used to identify gene products. Use a concentration of DNA that gives a big difference in gene products level with and without a repressor or an inducer present.
4. The DNA used should be very pure and free of RNA and proteins. It should be purified by CsCl gradient centrifugation. DNA from lambda phages and pBR322 derived plasmids have mainly been used in this system, but DNA from other sources can be used as well.
5. The magnesium concentration has a big influence on the system and should be optimized for each S-30 extract. Remember that the S-30 extracts also contain magnesium ions.
6. The easiest way of detecting a gene product in this system is to use a specific assay for the protein. If this is not possible, then use radioactive amino acids to detect the gene products.

## References


1. Zubay, G., (1973) *Ann. Rev. Gen.* **7,** 267–287.
2. Valentin-Hansen, P., Hammer-Jespersen, K., and Buxton, R. S., (1979) *J. Mol. Biol.* **133,** 1–17.
3. Miller, J. H. (1972) *Experiments in Molecular Genetics.* Cold Spring Harbor Laboratory, New York.

# Chapter 20

# In Vitro Translation of Messenger RNA in a Wheat Germ Extract Cell-Free System

***C. L. Olliver, A. Grobler-Rabie, and C. D. Boyd***

*MRC Unit for Molecular and Cellular Cardiology, University of Stellenbosch Medical School, Tygerberg, South Africa*

## Introduction

The wheat germ extract in vitro translation system has been used widely for faithful and efficient translation of viral and eukaryotic messenger RNAs in a heterologous cell-free system (*1–9*). With respect to the yield of translation products, the wheat germ extract is less efficient than most reticulocyte lysate cell-free systems. There are advantages however of using wheat germ extracts. Firstly, the in vivo competition of mRNAs for translation is more accurately represented, making the wheat germ system preferable for studying regulation of translation (*1*). Secondly, particularly low levels of endogenous mRNA and the endogenous nuclease activity (*14*) obviate the requirement for treatment with a calcium-activated nuclease. There is therefore less disruption of the in vivo situation and contamination with calcium ions is less harmful. The identification of all sizes of exogenous mRNA-directed translation products is facilitated because of the low levels of endogenous mRNA present. Thirdly, there is no post-translational modification of translation products; primary products are therefore investigated, although processing may be achieved by the addition of microsomal membranes to the translation reaction. Fourthly, the ionic conditions of the reaction may be altered to optimize the translation of large or small RNAs (*2*) (*see* Note 1). Translational activity is optimized by the incorporation of an energy-generating system of ATP, GTP, creatine phosphate, and creatine kinase (*3*). Wheat germ is inexpensive and commercially available (*see* Note 2); preparation of the extract is rapid and simple, resulting in high yields. Wheat germ extract cell-free system kits are also commercially available.

## Materials

Components of the wheat germ in vitro translation system are heat-labile and must be stored in aliquots of convenient volumes at −70°C. Freeze-thaw cycles must be minimized. Sterile techniques are used throughout. RNAse contamination is prevented by heat-sterilization (250°C, 8 h) of glassware and tips, and so on, or by diethyl pyrocarbonate treatment of glassware, followed by thorough rinsing of equipment in sterile distilled water.

1. Wheat Germ Extract. This is prepared essentially as described by Roberts and Paterson (*4*). The procedure must be carried out at 4°C, preferably in plastic containers since initiation factors stick to glass. Fresh wheat germ (approximately 5 g) (*see* Note 2) is ground with an equal weight of sand and 28 mL of 20 m*M* Hepes (pH 7.6), 100 m*M* KCl, 1 m*M* magnesium acetate, 2 m*M* $CaCl_2$, 6 m*M* 2-mercapteothanol, added

gradually. This mixture is then centrifuged at 28,000*g* for 10 min at 2°C, pH 6.5. This pH prevents the release of endogenous mRNA from polysomes and therefore removes the requirement for a pre-incubation to allow polysome formation (*4, 5*). The supernatant (S-28) is then separated from endogenous amino acids and plant pigments that are inhibitory to translation, by chromatography through Sephadex G-25 (coarse) in 20 m*M* Hepes (pH 7.6), 120 m*M* KCl, 5 m*M* magnesium acetate and 6 m*M* 2-mercaptoethanol. Reverse chromatography will prevent the loss of amino acids. Fractions of more than 20 $A_{260}$ nm/mL are pooled before being stored in aliquots at a concentration of approximately 100 $A_{260}$ nm/mL, at −70°C. The extract remains translationally active for a year or more.

2. L-[$^3$H]- or L-[$^{35}$S]-Amino Acids. 10–50 µCi of an appropriate amino acid [abundant in the protein(s) of interest] is added to the reaction to allow detection of translation products. Convenient specific activities are 140 Ci/mmol tritiated, or 1 Ci/mmol [$^{35}$S]-amino acids, respectively (*see* Note 3). Aqueous solutions should be used since ethanol, salts, detergents, and various solvents interfere with translation. Ethanol should be removed by lyophilization and the effects on translation of other solutions should be determined prior to their use. [$^{35}$S]-labeled amino acids must be stored in small aliquots at −70°C where they remain stable for up to six months, after which time sulfoxide products of degradation inhibit translation.
3. Messenger RNA. The extraction of both total and polyadenylated RNA has been described by a number of authors (*10–12*) (*see* Chapters 13–17). 1.5 mg/mL total RNA or 150 µg/mL polyadenylated RNA (in sterile distilled water) are convenient stock concentrations. RNA is stable for more than a year at −70°C. Contamination with potassium (*see* Note 1), phenol and ethanol must be prevented by 70% (v/v) ethanol washes, chloroform:butanol (4:1) extractions, and lyophilization respectively.
4. 10 × Energy Mix: 10 m*M* ATP, 200 µ*M* GTP, 80 m*M* creatine phosphate. Potassium salts of the nucleotide triphosphates should be used and the final pH ad-

justed (if necessary) to 7.4–7.6 with sodium hydroxide.

5. 0.5–1.0*M* potassium acetate (*see* Note 1), 25 m*M* magnesium acetate.
6. 20 m*M* dithiothreitol.
7. 0.6–1.2 m*M* spermine or 4.0–8.0 m*M* spermidine (*see* Note 4).
8. 0.2*M* Hepes (pH 7.4–7.6) (*see* Note 5).
9. 200–500 μg/mL creatine kinase (*see* Note 6).

# Method

All preparations are carried out on ice. After use, components are quick-frozen on dry ice. Reactions are carried out in sterile plastic microfuge tubes.

1. Mix the following solutions:

| *Component* | *Vol/50 μL reaction* |
|---|---|
| Energy mix | 5 μL |
| Potassium and magnesium acetate | 5 μL |
| Dithiothreitol | 5 μL |
| HEPES | 5 μL |
| Spermine | 5 μL |
| 0.3–8.0 μg mRNA, $dH_2O$ | 10 μL |
| Wheat germ extract | 10 μL (0.8–1.0 $A_{260}$ units) |
| Creatine kinase | 5 μL |

If a number of incubations are to be made, a master mix of the first five solutions may be prepared and 25 μL aliquoted/reaction tube. Creatine kinase is added last to ensure that no energy is wasted. The solutions are mixed by tapping the tube or by gentle vortexing. 0.5 $A_{260}$ units of microsomal membranes may be added before the creatine kinase to detect cotranslational modification of translation products (*see* Note 10).

2. Incubate at 28°C for 1 h (*see* Note 7). The reaction is terminated by placing the tubes at 4°C.

3. Incorporation of radioactive amino acids into mRNA-derived translation products is detected by TCA-precipitation of an aliquot of the reaction (*see* Chapter 21, Method 5 for procedure). Incorporation of radioactivity into translation products is generally not as well-stimulated by mRNA added to wheat germ extracts as it is in described reticulocyte lysates.
4. The remaining in vitro translation products may be analyzed further by standard techniques including tryptic mapping and ion-exchange chromatography, but specific products may be analyzed by immunoprecipitation followed by SDS-polyacrylamide gel electrophoresis (*see* Chapter 22).

## Notes

1. Wheat germ extract translational activity is particularly sensitive to variation in the concentration of potassium ions. At concentrations lower than 70 m*M*, small mRNAs are preferentially translated, whereas larger mRNAs are completely translated at potassium acetate concentrations of 70 m*M* or greater (*2*,*5*). Polypeptides of up to 200,000 daltons are synthesised under correct ionic conditions (*9*). Furthermore, chloride ions appear to inhibit translation such that potassium acetate should preferably be used (*5*). In this context, residual potassium should be removed from RNA preparations, by 70% (v/v) ethanol washes.
2. Inherent translational activity varies with the batch of wheat germ. Israeli mills (for example "Bar-Rav" Mill, Tel Aviv) supply wheat germ, the extracts of which are usually active.
3. Most of the endogenous amino acids have been removed by chromatography through Sephadex G-25 (coarse). Depending on the batch of wheat germ extract, addition of amino acids (to 25 μ*M*) and/or tRNA (to 58 μg/mL) may be necessary to optimize translational activity. Wheat germ extract is particularly sensitive to amino acid starvation; use of radioactive amino acids at specific activities greater than those suggested (*see* Materials, point 2) may result in inhibition of translation due to amino acid starvation.

4. The use of either spermine or spermidine generally stimulates translation, but is essential for the synthesis of larger polypeptides (*5*), probably by stabilizing longer mRNAs. Omission of either compound will increase the optimum magnesium acetate concentration to 4.0–4.3 m*M*.
5. HEPES has been shown to buffer the wheat germ extract in vitro translation system more effectively than Tris-acetate (*4*). Use of the latter will alter the optimum potassium and magnesium concentration.
6. Commercial preparations of creatine kinase differ with respect to the levels of nuclease contamination. This must be considered when larger amounts of the enzyme are to be used.
7. mRNA-stimulated incorporation of radioactive amino acids into translation products is linear, after a 5 min lag, for 50 min and is complete after 90 min. The system is labile at temperatures greater than 30°C; optimum activity is achieved at 25–30°C depending on the batch of wheat germ extract. An incubation temperature of 28°C is generally used.
8. In order to obtain maximum translational activity, it is necessary to determine the optima for the following for each preparation of wheat germ extract; mRNA concentration, potassium and magnesium concentrations, and incubation temperature. Take into account the concentration of salts in the wheat germ extract column eluate.
9. Heating of large mRNAs at 70°C for 1 min followed by rapid cooling on ice increases the efficiency of their translation in wheat germ extract in vitro translation systems.
10. Cotranslational processing of translation products may be detected by the addition of dog pancreas microsomal membranes to the translation incubation. They may be prepared as described by Jackson and Blobel (*11*) or may be ordered with a commercial translation kit. Microsomal membranes should be stored in aliquots of approximately 5 $A_{260nm}$ units in 20 m*M* Hepes (pH 7.5) at −70°C. Repeated freezing and thawing must be avoided.

# References


1. Steward, A. G., Lloyd, M., and Arnstein, H. R. V. (1977) Maintenance of the ratio of α and β globin synthesis in rabbit reticulocytes. *Eur. J. Biochem.* **80,** 453–459.
2. Benveniste, K., Wilczek, J., Ruggieri, A., and Stern, R. (1976) Translation of collagen messenger RNA in a cell-free system derived from wheat germ. *Biochem.* **15,** 830–835.
3. Huntner, A. R., Farrell, P. J., Jackson, R. J., and Hunt, T. (1977) The role of polyamines in cell-free protein synthesis in the wheat germ system. *Eur. J. Biochem.* **75,** 149–157.
4. Roberts, B. E., and Paterson, B. M. (1973) Efficient translation of tobacco mosaic virus RNA and rabbit globin 9S RNA in a cell-free system from commercial wheat germ. *Proc. Natl. Acad. Sci. USA* **70,** 2330–2334.
5. Davies, J. W., Aalbers, A. M. J., Stuik, E. J., and van Kammen, A. (1977) Translation of cowpea mosaic virus RNA in cell-free extract from wheat germ. *FEBS Lett* **77,** 265–269.
6. Boedtker, H., Frischauf, A. M., and Lehrach, H. (1976) Isolation and translation of calvaria procollagen messenger ribonucleic acids. *Biochem.* **15,** 4765–4770.
7. Patrinou-Georgoulas, M., and John, H. A. (1977) The genes and mRNA coding for the theory chains of chick embryonic skeletal myosin. *Cell* **12,** 491–499.
8. Larkins, B. A., Jones, R. A., and Tsai, C. Y. (1976) Isolation and in vitro translation of zein messenger ribonucleic acid. *Biochem.* **15,** 5506–5511.
9. Schröder, J., Betz, B., and Hahlbrock, K. (1976) Light-induced enzyme synthesis in cell suspension cultures of *petroselinum*. *Eur. J. Biochem.* **67,** 527–541.
10. Adams, S. L., Sobel, M. E., Howard, B. H., Olden, K., Yamada, K. M., De Crombrugghe, B., and Pastan, I. (1977) Levels of translatable mRNAs for cell-surface protein, collagen precursors, and two membrane proteins are altered in Rous sarcoma virus-transformed chick embryo fibroblasts. *Proc. Natl. Acad. Sci. USA* **74,** 3399–3403.
11. Darnbrough, C. H., Legon, S., Hunt, T., and Jackson, R. J. (1973) Initiation of protein synthesis: Evidence for messenger RNA-independent binding of methionyl-transfer RNA to the 40S ribosomal subunit. *J. Mol. Biol.* **76,** 379–403.
12. Pemberton, R. E., Liberti, P., and Baglioni, C. (1975) Isolation of messenger RNA from polysomes by chromatography on oligo d(T) cellulose. *Anal. Biochem.* **66,** 18–28.

13. Jackson, R. C., and Blobel, G. (1977) Post-translational cleavage of presecretory proteins with an extract of rough microsomes, from dog pancreas, with signal peptidase activity. *Proc. Natl. Acad. Sci. USA* **74,** 5598–5602.
14. Pelham, H. R. B., and Jackson, R. J. (1976) An efficient mRNA-dependent translation system from reticulocyte lysates. *Eur. J. Biochem.* **67,** 247–256.

# Chapter 21

# In Vitro Translation of Messenger RNA in a Rabbit Reticulocyte Lysate Cell-Free System

***C. L. Olliver and C. D. Boyd***

*MRC Unit for Molecular and Cellular Cardiology, University of Stellenbosch Medical School, Tygerberg, South Africa*

## Introduction

The identification of specific messenger RNA molecules and the characterization of the proteins encoded by them, has been greatly assisted by the development of in vitro translation systems. These cell-free extracts comprise the cellular components necessary for protein synthesis, i.e., ribosomes, tRNA, rRNA, amino acids, initiation, elongation and termination factors, and the energy-generating system (*1*). Heterologous mRNAs are faithfully and efficiently translated in extracts of HeLa cells (*2*), Krebs II ascites tumor cells (*2*), mouse L cells (*2*), rat and mouse liver cells (*3*), Chinese hamster ovary (CHO) cells (*2*), and rabbit reticulocyte lysates (*2,4*), in addition to those of rye embryo (*5*) and wheat germ (*6*). Translation in cell-free systems is simpler and more rapid (60 min vs 24 h) than the in vivo translation system using Xenopus oocytes.

The synthesis of mRNA translation products is detected by their incorporation of radioactively labeled amino acids, chosen specifically to be those occurring in abundance in the proteins of interest. Analysis of translation products usually involves specific immunoprecipitation (*7*), followed by polyacrylamide gel electrophoresis (*8*) and fluorography (*9*) (*see* Fig. 1 and Chapter 22).

In vitro translation systems have played important roles in the identification of mRNA species and the characterization of their products, the investigation of transcriptional and translational control and the cotranslational processing of secreted proteins by microsomal membranes added to the translation reaction (*10, 11*). This chapter describes the rabbit reticulocyte lysate system for in vitro translation of mRNA.

Although the endogenous level of mRNA is low in reticulocyte lysates, it may be further reduced in order to maximize the dependence of translation on the addition of exogenous mRNA. This reduction is achieved by treatment with a calcium-activated nuclease that is thereafter inactivated by the addition of EGTA (*4*). The system is thus somewhat disrupted with respect to the in vivo situation and is particularly sensitive to the presence of calcium ions. The resulting lysate, however, is the most efficient in vitro translation system with respect to the exogenous mRNA-stimulated incorporation of radioactive amino acids into translation products. It is therefore particularly appropriate for the study of translation products. The system is sensitive, however, to regulation by a number of factors, including hemin, double-stranded RNA, and depletion of certain metabolites. The effects of these factors on regulation of translation of various mRNAs may therefore be investigated. Despite the efficiency of reticulocyte lysates, the competition for initiation of translation by various mRNA species may differ from the in vivo situation. Products therefore may not be synthesized at in vivo

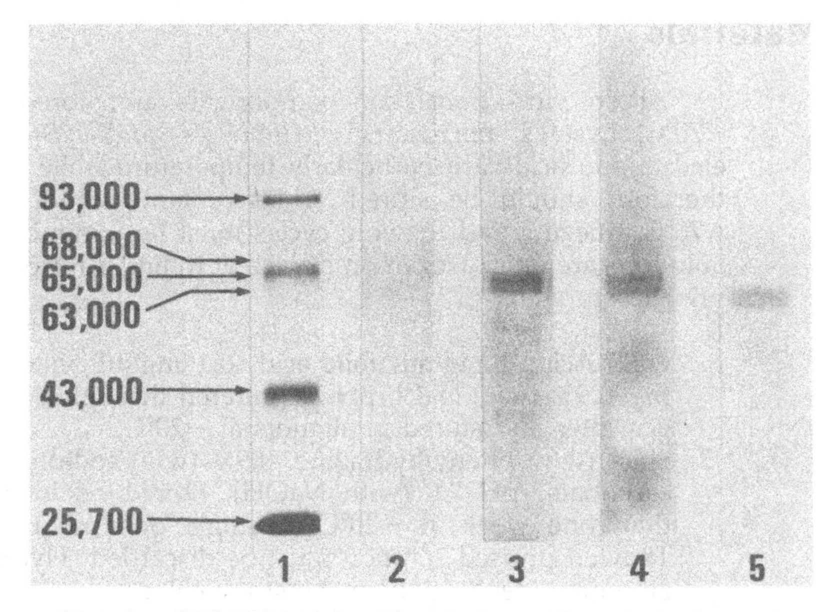


Fig. 1. SDS Polyacrylamide gel electrophoretic analysis of in vitro translation products. In vitro translation products were derived from exogenous mRNA in an mRNA-dependent reticulocyte lysate cell-free system. Following electrophoresis on 8% SDS polyacrylamide gels, radioactive protein products were analyzed by fluorography. Lane 1: [$^{14}$C]-labeled proteins of known molecular weights, i.e., phosphorylase A (93K), bovine serum albumin (68K), ovalbumin (43K), α-chymotrypsinogen (25.7K). Lanes 2–5 represent [$^{3}$H]-proline-labeled translation products of the following mRNAs: Lane 2: endogenous reticulocyte lysate mRNA, Lane 3: 0.3 μg calf nuchal ligament polyadenylated RNA. Lane 4: 0.3 μg calf nuchal ligament polyadenylated RNA, and immunoprecipitated with 5 μL sheep antiserum raised to human tropoelastin, Lane 5: 0.3 μg calf nuchal ligament polyadenylated RNA and cotranslationally processed by 0.3 $A_{260}$ nm microsomal membranes.

proportions; the wheat germ extract cell-free system (*see* Chapter 20) reflects the in vivo situation more faithfully. Nuclease-treated rabbit reticulocyte lysate cell-free systems are available as kits from a number of commercial suppliers.

## Materials

All in vitro translation components are stored at −70°C. Lysates, microsomal membranes, and [$^{35}$S]-labeled amino acids are particularly temperature-labile and therefore should be stored in convenient aliquots at −70°C; freezing and thawing cycles must be minimized. Solutions are quick-frozen on dry ice or in liquid nitrogen prior to storage.

1. Folic Acid; 1 mg/mL folic acid, 0.1 mg/mL vitamin $B_{12}$, 0.9% (w/v) NaCl, pH 7.0; filtered through a 0.45 µm filter and stored in aliquots at −20°C.
2. 2.5% (w/v) phenylhydrazine, 0.9% (w/v) sodium bicarbonate, pH 7.0 (with NaOH). Stored no longer than one week at −20°C in single dose aliquots. Thawed unused solution must be discarded. Hydrazine degrades to darken the straw color.
3. Physiological saline: 0.14*M* NaCl, 1.5 m*M* $MgCl_2$, 5 m*M* KCl. Stored at 4°C.
4. 1 m*M* hemin.
5. 0.1*M* $CaCl_2$.
6. 7500 U/mL micrococcal nuclease in sterile distilled water. Stored at −20°C.
7. Rabbit Reticulocyte Lysate. This is prepared essentially as described by Pelham and Jackson (*4*). Rabbits are made anemic by intramuscular injection of 1 mL folic acid solution on day one, followed by six daily injections of 0.25 mL/kg body weight of 2.5% phenylhydrazine solution (*see* Note 1). At a reticulocyte count of at least 80%, blood is collected on day 7 or 8 by cardiac puncture into a 200-mL centrifuge tube containing approximately 3000 units of heparin, and mixed well. Preparation should continue at 2–4°C.

   (a) Blood is centrifuged at 120 g, 12 min., 2°C, and plasma removed by aspiration.
   (b) Cells are resuspended in 150 mL *ice cold* saline and washed at 650*g* for 5 min. Washing is repeated three times.

(c) The final pellets are rotated gently in the bottle, then transferred to Corex tubes (which are only half-filled). An equal volume of saline is added, the cells gently suspended, then pelleted at 1020*g* for 15 min at 2°C. The leukocytes (buffy coat) are then removed by aspiration with a vacuum pump.

(d) In an ice bath, an equal volume of ice-cold sterile deionized distilled water is added and the cells lysed by vigorous vortexing for 30 s. (*See* Note 2). The suspension is then immediately centrifuged at 16,000*g* for 18 min at 2°C.

(e) At 4°C the supernatant is carefully removed from the pellet of membranes and cell debris. This lysate is then quick frozen in liquid nitrogen in aliquots of approximately 0.5 mL.

8. The optimum hemin concentration is determined by varying its concentration from 0 to 1000 μ*M* during the micrococcal nuclease digestion. 477.5 μL lysate, 5 μL 0.1*M* $CaCl_2$, 5 μL nuclease (75 U/mL final concentration) is mixed. A 97.5 μL volume of this is incubated with 2.5 μL of the relevant hemin concentration at 20°C for 20 min. A 4 μL 0.05*M* solution of EGTA is added to stop the digestion. The optimum hemin concentration is that allowing the greatest translational activity (incorporation of radioactive amino acids) in a standard cell-free incubation (*see* Method section). A 25 μ*M* quantity is generally used to ensure efficient chain initiation.
   Lysates are extremely sensitive to ethanol, detergents, metals, and salts, particularly calcium. Stored at −70°C, reticulocyte lysates remain active for more than 6 months.
9. L-[$^3$H]- or L-[$^{35}$S]-Amino Acids. A radioactive amino acid, labeled to a high specific activity (140 Ci/mmol tritiated, or approximately 1 Ci/mmol [$^{35}$S]-labeled amino acids), is added to the translation incubation to enable detection of the translation products. An amino acid known to be abundant in the protein of interest is chosen. Radioactive solutions should prefera-

bly be aqueous; those of low pH should be neutralized with NaOH; ethanol should be removed by lyophilization, and the effect of solvents on lysate activity should be tested. [$^{35}$S] degrades rapidly to sulfoxide and should be aliquoted and stored at −70°C to prevent interference by sulfoxides.

10. Messenger RNA. Total RNA may be extracted from various tissues by a number of methods (*see* Chapters 13–17). RNA stored in sterile $dH_20$ at −70°C is stable for more than a year. Contamination by ions, metals, and detergents should be avoided.
    Phenol may be removed by chloroform:butanol (4:1) extractions; salts are removed by precipitation of RNA in 0.4*M* potassium acetate (pH 6.0) in ethanol. Ethanol should be removed by lyophilization. Convenient stock concentrations for translation are 1.5 mg/mL total RNA or 150 μg/mL $polyA^+$ RNA.
11. Translation cocktail: 250 m*M* Hepes (pH 7.2), 400 m*M* KCl, 19 amino acids at 500 m*M* each (excluding the radioactive amino acid), 100 m*M* creatine phosphate.
12. 20 m*M* magnesium acetate (pH 7.2).
13. 2.0*M* potassium acetate (pH 7.2).
14. Sterile distilled $H_20$.

Sterile techniques are used; RNAse contamination is avoided by heat-treatment of glassware (250°C, 12 h) or by treatment of heat-sensitive materials with diethylpyrocarbonate, followed by rinsing in distilled water. Sterile gloves are worn throughout the procedure.

## Method

In vitro translation procedures are best carried out in autoclaved plastic microfuge tubes (Eppendorf); a dry incubator is preferable to waterbaths for provision of a constant temperature. All preparations are performed on ice.

1. A reaction mix of the following constituents is prepared on ice:

| *Component* | *µL/incubation* | *µL/10 incubations* |
|---|---|---|
| $dH_2O$ | 0.7 | 7 |
| 2.0*M* potassium acetate | 1.3 | 13 |
| 10–50 µCi radioactive amino acid | 5 | 50 |
| Translation cocktail | 3 | 30 |

Components are added in the above order, vortexed, and 10 µL is aliquoted per incubation tube on ice.

2. mRNA and $dH_20$ in l0 µL is added (*see* Note 5). For example, 8 µL $dH_20$ is added to 2 µL 1.5 mg/mL total RNA. A control incubation without the addition of exogenous mRNA detects translation products of residual endogenous reticulocyte mRNA.
3. A 10 µL volume of lysate is added last to initiate translation. If required, 0.5 $A_{260}$ nm units of microsomal membranes are also added at this point for cotranslational processing of translation products (*see* Note 10).
4. The mixture is vortexed gently prior to incubation at 37°C for 60 min. The reaction is stopped by placing the tubes on ice.
5. Detection of mRNA-directed incorporation of radioactive amino acids into translation products is performed by determination of acid-precipitable counts.

   At the initiation and termination of the incubation, 5 µL aliquots are spotted onto glass fiber filters that are then air-dried. Filters are then placed into 10 mL/filter of the following solutions:

   (i) 10% (v/v) cold trichloroacetic acid (TCA) for 10 min on ice.
   (ii) 5% (v/v) boiling TCA for 15 min, to degrade primed tRNAs.
   (iii) 5% (v/v) cold TCA for 10 min on ice.

   The filters are then washed in 95% (v/v) ethanol, then in 50% (v/v) ethanol–50% (v/v) acetone, and finally in 100% (v/v) acetone. The filters are dried at 80°C for 30 min. TCA-precipitated radioactivity is determined by immersing the filters in 5 mL toluene-based scintillation fluid and counting in a scintillation counter.

Exogenous mRNA-stimulated translation can be expected to result in a five- to 30-fold increase over background of incorporation of [$^3$H]- or [$^{35}$S]-labeled amino acids, respectively, into translation products.

6. An equal volume of 2% (w/v) SDS, 20% (w/v) glycerol, 0.02% (w/v) bromophenol blue, 1*M* urea is added to the remaining 20 μL of translation mixture. This is made 0.1*M* with respect to dithiothreitol, heated at 95°C for 6 min, and slowly cooled to room temperature prior to loading onto a polyacrylamide gel of appropriate concentration (between 6 and 17%). After electrophoresis, radioactive areas of the gel are visualized by fluorography (Fig. 1).

# Notes

1. Maximum anemia may be achieved by reducing the dose of phenylhydrazine on day 3, then increasing it on following days. The reticulocyte count is determined as follows:
   (i) 100 μL blood is collected in 20 μL of 0.1% heparin in saline.
   (ii) 50 μL blood heparin is incubated at 37°C for 20 min with 50 μL of 1% (w/v) brilliant cresyl blue, 0.6% (w/v) sodium citrate, 0.7% (w/v) sodium chloride.
   (iii) Reticulocytes appear under the microscope as large, round, and with blue granules. Erythrocytes are small, oval, and agranular.
2. The volume of water (in mL) required to lyse the reticulocyte preparation is equal to the weight of the pellet in the tube.
3. Endogenous mRNAs of lysates are degraded by a calcium-activated nuclease that is inactivated by

EGTA. Lysates are therefore sensitive to calcium ions, the addition of which must be avoided to prevent degradation of added mRNAs by this activated nuclease.

4. Vigorous vortexing decreases efficiency of translation, therefore do so gently when preparing the reaction mix.
5. The optimum mRNA concentration should be determined prior to definitive experiments by varying the mRNA concentrations while keeping other variables constant. Care should be taken to avoid excess mRNA; polyadenylated RNA in excess of 1 μg has been noted to inhibit translation.
6. Heating of mRNA at 70–80°C for 1 min followed by quick cooling in an ice bath, prior to addition to the incubation mixture, has been shown to increase the efficiency of translation of GC-rich mRNA; for example, heating elastin mRNA at 70–80°C prior to translation resulted in a 100% increase, compared with unheated mRNA, of incorporation of radioactivity into translation products (*12*).
7. Optimum potassium concentrations may vary from 30 to 100 m*M* depending on mRNAs used and should be determined prior to definitive translations. Similarly, specific mRNAs may require altered magnesium concentrations, although a concentration of 0.6–1.0 m*M* is generally used.
8. The addition of spermidine at approximately 0.4 m*M* has been noted to increase translation efficiency in certain cases (*12*), possibly by stabilizing relevant nucleic acids. However, this effect may also be lysate-dependent and should be optimized if necessary for individual lysate preparations.
9. Specific activities greater than those mentioned (Materials, item 2) may result in depletion of the amino acid concerned, with subsequent inhibition of translation.
10. Cotranslational processing of translation products may be detected by the addition of dog pancreas microsomal membranes to the translation incubation. These may be prepared as described by Jackson and Blobel (*11*) or may be ordered with a commercial translation kit. Microsomal membranes should be

stored in aliquots of approximately 5 $A_{260}$ nm units in 20 m*M* Hepes (pH 7.5) at −70°C. Repeated freezing and thawing must be avoided.

## References


1. Lodish, H. F. (1976) Translational control of protein synthesis. *Ann. Rev. Biochem.* **45,** 39–72.
2. McDowell, M. J., Joklik, W. K., Villa-Komaroff, L., and Lodish, H. F. (1972) Translation of reovirus messenger RNAs synthesized in vitro into reovirus polypeptides by several mammalian cell-free extracts. *Proc. Natl. Acad. Sci. USA* **69,** 2649–2653.
3. Sampson, J., Mathews, M. B., Osborn, M., and Borghetti, A. F. (1972) Hemoglobin messenger ribonucleic acid translation in cell-free systems from rat and mouse liver and Landschutz ascites cells. *Biochem.* **11,** 3636–3640.
4. Pelham, H. R. B., and Jackson, R. J. (1976) An efficient mRNA-dependent translation system from reticulocyte lysates. *Eur. J. Biochem.* **67,** 247–256.
5. Carlier, A. R., and Peumans, W. J. (1976) The rye embryo system as an alternative to the wheat-system for protein synthesis in vitro. *Biochem. Biophys. Acta* **447,** 436–444.
6. Roberts, B. E., and Paterson, B. M. (1973) Efficient translation of tobacco mosaic virus RNA and rabbit globin 9S RNA in a cell-free system from commercial wheat germ. *Proc. Natl. Acad. Sci. USA* **70,** 2330–2334.
7. Kessler, S. W. (1981) Use of protein A-bearing staphylococci for the immunoprecipitation and isolation of antigens from cells. In *Methods in Enzymology* (Langone, J. J., and Van Vunakis, H., eds.) **73,** 441–459. Academic Press, New York.
8. Laemmli, U. K. (1970) Cleavage of structural proteins during the assembly of the head of bacteriophage T4. *Nature* **227,** 680–685.
9. Bonner, W. M., and Laskey, R. A. (1974) A film detection method for tritium-labelled proteins and nucleic acids in polyacrylamide gels. *Eur. J. Biochem.* **46,** 83–88.
10. Shields, D., and Blobel, G. (1978) Efficient cleavage and segregation of nascent presecretory proteins in a reticulocyte lysate supplemented with microsomal membranes. *J. Biol. Chem.* **253,** 3753–3756.

11. Jackson, R. C., and Blobel, G. (1977) Post-translational cleavage of presecretory proteins with an extract of rough microsomes, from dog pancreas, with signal peptidase activity. *Proc. Natl. Acad. Sci. USA* **74,** 5598–5602.
12. Karr, S. R., Rich, C. B., Foster, J. A., and Przybyla, A. (1981) Optimum conditions for cell-free synthesis of elastin. *Coll. Res.* **1,** 73–81.

# Chapter 22

# Immunoprecipitation of In Vitro Translation Products with Protein A Bound to Sepharose

**_C. L. Olliver and C. D. Boyd_**

*MRC Unit for Molecular and Cellular Cardiology, University of Stellenbosch Medical School, Tygerberg, South Africa*

## Introduction

The entire complement of in vitro translation products derived from a mRNA population may be analyzed by polyacrylamide gel electrophoresis followed by fluorography and autoradiography. It is often necessary however to demonstrate the synthesis of a polypeptide translation product present in minimal amounts, or one comigrating with another product. In such cases, and also to prove the identity of particular translation products, it is possible to separate the particular polypeptide from the general population by specific immunoprecipitation with antisera raised specifically to the polypeptide of interest.

This method involves initial complexing of antibodies with the relevant antigens. Protein A (isolated from the cell walls of *Staphylococcus aureus*) then binds to the constant regions of the immunoglobins (*1*). This antigen–antibody–protein A complex may be precipitated by virtue of the Sepharose attached to protein A. In this manner, quantitative considerations regarding antigen and first antibody concentration ratios are avoided, as are those relating to first and second antibodies, normally essential for precipitation of such immune complexes. In addition, this procedure is much faster and more specific than the double antibody procedure. In order to analyze immunoprecipitates by gel electrophoresis instead of by simple dpm determinations, the protein A–Sepharose may be easily dissociated from the immune complex by heating. Specifically immunoprecipitated proteins may then be analyzed electrophoretically.

## Materials

1. NET. 150 m*M* NaCl, 5 m*M* EDTA, 50 m*M* Tris-HCl (pH 7.4).
2. 2% SDS, 5 m*M* dithiothreitol, 10 m*M* Tris-HCl (pH 8.3).
3. NP-40.
4. Antiserum raised specifically to the in vitro translation product under investigation should be stored at 4 or −70°C.
5. 0.1 g Protein A bound to Sepharose (Pharmacia Fine Chemicals, Inc.) is swollen for 30 min at room temperature in 1 mL 10% (w/v) sucrose, 0.5% NP-40 then washed three times in 1 mL 0.05% (v/v) NP-40 in NET, twice in NET, then suspended in 3.5 mL NET, to be stored at 4°C. Since this is a suspension, the protein A-Sepharose tends to settle and must be mixed immediately prior to use.
6. In vitro translation reaction mixtures may be stored at −20°C for up to a week if they are not to be immunoprecipitated immediately following incubation.

# Method

1. 25 μL NET is added to 25 μL of in vitro translation incubation mixture.
2. After addition of 5 μL antiserum, the mixture is incubated at 4°C for 2 h with occasional gentle mixing, then at 20°C for 15 min with constant shaking.
3. 50 μL protein A-Sepharose suspension is then added and incubation continues at 20°C for 45 min with constant slow shaking.
4. Immune complexes are precipitated for 5 min in an Eppendorf microfuge, then washed twice with 50 μL NET with 3 min precipitations in a microfuge.
5. Final immune-complex pellets are dissolved in 50 μL 2% SDS, 5 m*M* dithiothreitol, 10 m*M* Tris-HCl, pH 8.3.
6. Protein A-Sepharose is dissociated from the antigen–antibody complex by boiling for 2 min, then precipitating for 5 min in a microfuge.
7. TCA-precipitable counts are determined as previously described on 5 μL aliquots of the final supernatants (which contain the immune complexes). The remainder may be applied to a polyacrylamide gel of appropriate concentration for visual analysis by fluorography (*see* Chapter 17, Vol. 1).

# Notes

1. The initial antigen–antibody incubation may be reduced to 1 h without significant reduction in yield of immunoprecipitate. Similarly, the incubation with protein A-Sepharose may be reduced to 15 min; however, long incubations ensure complete binding and precipitation.
2. Residual protein A-Sepharose should not enter 8–17% polyacrylamide gels; however, easier sample loading is facilitated by optimum separation from the immune complex after precipitation, i.e., by carefully drawing off the supernatant with an automatic pipet.
3. The amount of antiserum required to immunoprecipitate the translation products will vary according to

the antibody affinity and avidity; a series of volumes of antiserum added to the lysate will allow determination of the optimum amount required for maximum immunoprecipitation of the translation products.

## References

1. Kessler, S. W. (1981) Use of protein A-bearing staphylococci for the immunoprecipitation and isolation of antigens from cells. In *Methods in Enzymology* (Langone, J. J., and Van Vunakis, H., eds.) **73,** 441–459. Academic Press, New York.

# Chapter 23

# In Vitro Continuation of RNA Synthesis Initiated In Vivo

***Theodore Gurney, Jr.***

*Department of Biology, University of Utah, Salt Lake City, Utah*

## Introduction

Isolated nuclei will continue synthesis of RNA initiated in vivo, but reinitiation of synthesis is rare in washed nuclei (*1*). This situation can be exploited to measure instantaneous rates of in vivo transcription because the cell-free conditions are well-defined and nascent transcripts are generally not subject to the rapid cleavage often found in living cells (*2–5*). Isolated nuclei can also be used to map a primary transcript on genomic DNA. The method has been used to show that transcription terminates more than 1000 nucleotides downstream from the poly A site in β-major globin mRNA (*6*).

Washed nuclei are incubated in a reaction mix containing radioactive nucleoside triphosphates. The resulting radioactive RNA is then hybridized to an excess of cloned DNA homologous to the gene in question. Results are expressed as the fraction of total in vitro synthesized RNA that hybridizes to a particular clone. Large clones, spanning the gene, would be used to quantify gene-specific transcription (*2,5*). Small clones, subdividing the gene and the flanking sequences, would be used to map the ends of a primary transcript (*6*).

I describe here a method of in vitro RNA synthesis using crude nuclei isolated from cultured cells. Methods of hybridization, needed to complete the analysis, are described elsewhere (*2,6,7*).

## Materials

The procedures are described for cultured cells grown in monolayer. The same methods could probably be adapted to tissues or suspension cultures.

1. Dulbecco's modified Eagle's medium.
2. Calf serum.
3. Tissue culture grade Petri plates, 57 $cm^2$.
4. SVT2 mouse cells are cultured in 5% calf serum to a density of $3.5 \times 10^5$ cells/$cm^2$.
5. Phosphate buffered saline (PBS): 140 m*M* NaCl, 2.7 m*M* KCl, 8.1 m*M* $Na_2HPO_4$, 1.5 m*M* $KH_2PO_4$, 0.9 m*M* $CaCl_2$, 0.4 m*M* $MgCl_2$.

   The solution minus calcium and magnesium salts is autoclaved. A 100× stock solution of the calcium and magnesium salts is autoclaved separately and added later to the cold salt solution. PBS is stored at 4°C.
6. Lysis buffer: 30 m*M* Tris-HCl, pH 7.9; 80 m*M* KCl; 7 m*M* magnesium acetate; 7 m*M* 2-mercaptoethanol; 0.01 m*M* EDTA; 10% (v/v) glycerol; 0.5% Nonidet P-40 or Triton X-100.

   Lysis buffer is stored at 4°C and is stable for 2 months.
7. Reaction mix (without radioactive UTP): 30 m*M* Tris-HCl, pH 7.9; 80 m*M* KCl; 7 m*M* magnesium acetate; 7 m*M* 2-mercaptoethanol; 0.01 m*M* EDTA; 10% glycerol; 1 m*M* ATP [molar extinction coefficient (*8*) at 260

nm and pH 7: 15.4]; 0.4 m*M* CTP [molar extinction coefficient (*8*) at 260 nm and pH 7: 7.4]; 0.4 m*M* GTP [molar extinction coefficient (*8*) at 260 nm and pH 7: 11.7]; 0.0–0.1 m*M* UTP [molar extinction coefficient (*8*) at 260 nm and pH 7: 9.9].

The mix is stored in quantities of 0.5 mL at −20°C. It is stable for 1 yr.

8. 5-$^3$H-UTP or α-$^{32}$P UTP are purchased at high specific radioactivities in 50% aqueous ethanol and are stored at −20°C for as short a time as possible. A small quantity (typically 100 μCi) is evaporated to dryness at the time of the experiment and is mixed with reaction mix and the nuclei, as described below (*see* Note 1).
9. Stop mix (used to stop the reaction in studies of kinetics): 100 m*M* NaCl; 10 m*M* Tris-HCl, pH 7.5; 1 m*M* EDTA; 100 μg/mL commercial yeast RNA; 0.2% (w/v) sodium dodecyl sulfate. The stop mix is stored at room temperature.
10. Stop acid: 20% trichloroacetic acid (w/v) plus 20 m*M* sodium pyrophosphate. Store at 4°C.
11. Washing acid: 1*N* HCl plus 20 m*M* sodium pyrophosphate. Store at 4°C.
12. Reagents for purifying RNA are described in Chapter 3.

# Methods

1. Count a culture to estimate the number of cells. A typical experiment would use 2 × 10$^7$ isolated nuclei from one culture, incubated in a volume of 100 μL of reaction mix, containing 100 μCi of radioactive UTP. SV40-transformed mouse cells can easily attain 2 × 10$^7$ cells per culture in log-phase growth.
2. Evaporate 100 μCi of radioactive UTP to dryness. The UTP in 50% ethanol is placed in the bottom of a 12 × 77 mm polypropylene test tube that is placed upright in a small vacuum desiccator. A vacuum pump is attached to the desiccator, with a cold trap. Vacuum is applied, at first slowly to avoid explosive boiling. The drying of 100 μL should take about 20 min at 22°C. As

soon as the tube is dry, add 80 μL of reaction mix and store on ice.

3. Carry the warm culture(s) containing $2\times 10^7$ cells to the cold room. Decant the warm medium and immediately (<2 s) rinse the plate twice with about 30 mL of cold PBS, by pouring. Try to chill the cultures as rapidly as possible. After two rinses, drain the plate for 1 min and remove residual PBS with a Pasteur pipet, then cover the monolayer with 5 mL of lysis buffer. Wait for 2 min at 4°C, then decant the lysis buffer. Nuclei and insoluble structural proteins will remain attached to the plate.
4. Add 3 mL of fresh lysis buffer to the plate and detach the nuclei using vigorous pipeting up and down with a short pipet and bulb. Put the suspended material in a 12 × 77 mm polypropylene tube and centrifuge it (2000*g*, 2 min, 4°C).
5. Decant the supernatant, drain for 1 min, and resuspend the pellet in the radioactive reaction mix, on ice, by pipeting up and down with a micropipet. To avoid clumping of nuclei, the centrifugation should be as short as possible to give a rather loose pellet, and the resuspension should be thorough.
6. To determine the kinetics of synthesis, the suspension of nuclei is subdivided into samples of 5 μL. For isolation of the labeled RNA, the suspension is not subdivided.
7. The suspension is then incubated at 30°C, for 5–30 min.
8. To measure rates of incorporation of UTP into acid-insoluble material (presumptive RNA), incubate duplicate samples of 5 μL for 0, 5, 10, 20, and 30 min. At the end of incubation, add 0.2 mL of stop mix to the sample and vortex at room temperature. The material is now stable for hours.
9. Add 0.2 mL of 20% TCA, mix, and chill on ice for 60 min to precipitate nucleic acids.
10. Filter the precipitate onto a GF/C filter, rinsing 5 times with 1*N* HCl plus pyrophosphate, at 0°C. Rinse the filter once with a small amount of 95% ethanol.

11. Dry the filter under vacuum and count it in a toluene-based scintillation fluid, without solubilizer. The incorporation data may be related to nuclear DNA by using the fluorescence assay for DNA (Chapter 2).
12. To isolate the labeled RNA, incubate an undivided sample for 20 min, then add 5 mL of pronase-SDS. Vortex the mixture and incubate it at 40°C for 30 min. (*See* Chapter 3.)
13. Add an equal volume of phenol–chloroform and vortex hard again to give a uniform emulsion, at room temperature.
14. Break the emulsion by centrifugation (2000*g*, 2 min, 23°C).
15. Remove the upper aqueous phase with a Pasteur pipet into a weighed centrifuge tube. Weigh the tube again and add 95% ethanol, 2.5 times the weight of the aqueous phase. Mix thoroughly and chill at −20°C for at least 3 h.
16. Centrifuge (8000*g*, 30 min, 0°C) and decant the supernatant.
17. Rinse the tube by filling it half-full with cold 70% ethanol, mixing, centrifuging again (8000*g*, 5 min, 0°C), and decanting. Drain the tube in the cold for 5 min.
18. Concentration of the RNA is aided by degrading the nuclear DNA with RNase-free DNase, prepared as described in Chapter 3. Suspend the washed nucleic acids at 50 μg DNA/mL in DNase digestion buffer and add 1 Kunitz unit/mL of DNase. Incubate for 30 min at 4°C, add an equal volume of pronase-SDS and repeat the steps above in the isolation of RNA. The end product will be a washed precipitate of undegraded radioactive RNA and partially degraded cellular DNA.
19. The products of in vitro synthesis are shown in Fig. 1, as displayed by electrophoresis in a 0.7% agarose–formaldehyde gel. In 20 min of synthesis, the ribosomal RNA was largely completed to 14 kb, the size of the primary transcript. Nonribosomal RNA was labeled (in Lane a) in molecules both larger and smaller than 14 kb. The product size was stable in longer incu-

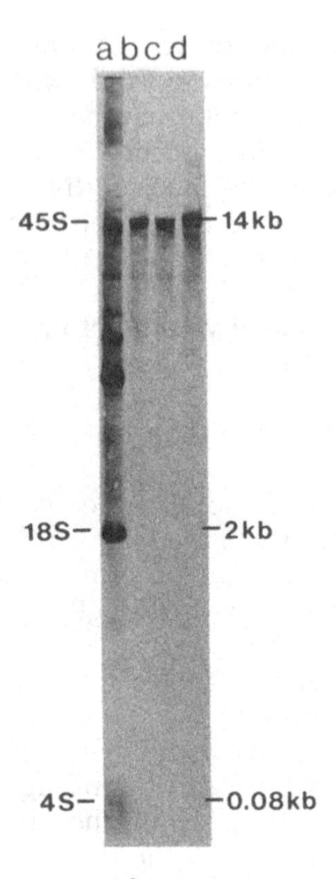


Fig. 1. Electrophoresis of $^{3}$H-labeled nucleic acids from SVT2 cells in a 0.7% agarose–formaldehyde gel. Cells were labeled either in vivo or in vitro as described. Lane a: SVT2 cells labeled in vivo 3 h with $^{3}$H-uridine. Lane b: SVT2 nuclei labeled in vitro 20 min with $^{3}$H-UTP plus 300 μg/mL α-amanitin, to confine synthesis to ribosomal RNA. Synthesis stops before 20 min at a level of 1.0 pmol UTP incorporated per μg DNA. Lane c: A second sample, prepared as for Lane b. Lane d: SVT2 nuclei labeled in vitro 20 min with $^{3}$H-UTP, no α-amanitin. Synthesis stops before 20 min at a level of 2.4 pmol UTP incorporated per μg DNA.

bation, in the absence of further incorporation. Very little incorporation was found in the small RNA region, 4S or 5S.

# Notes

1. The concentration of UTP is usually determined by its specific radioactivity. Tritiated UTP is available at up

to 50 Ci/mmol, which gives a UTP concentration of ≥20 μ*M* in the reaction. That concentration will allow extensive addition to an RNA molecule, several thousand nucleotides in 20 min. High specific activity $\alpha$-$^{32}$P-UTP is available at 3000 Ci/mmol and gives a UTP concentration of 300 n*M*. That concentration allows only a few nucleotides continuation, probably less than 100. The limited synthesis with high specific activity UTP is the method of choice for mapping the ends of a primary transcript, so as to confine heavy labeling to a small extension of ongoing synthesis.

## Acknowledgments

This work as supported by USPHS Grant GM 26137 and a grant from the University of Utah Research Committee.

## References

1. Udvardy, A., and Seifart, K. H. (1976) Transcription of specific genes in isolated nuclei from HeLa cells in vitro. *Eur. J. Biochem.* **62,** 353–363.
2. McKnight, G. S., and Palmiter, R. D. (1979) Transcriptional regulation of the ovalbumin and conalbumin genes by steroid hormones in chick oviduct. *J. Biol. Chem.* **254,** 9050–9058.
3. Hofer, E., and Darnell, J. E., Jr. (1981) The primary transcription unit of the mouse β-major globin gene. *Cell* **23,** 585–593.
4. Derman, E., Krauter, K., Walling, L., Weinberger, C., Ray, M., and Darnell, J. E., Jr. (1981) Transcriptional control of liver-specific mRNAs. *Cell* **23,** 731–739.
5. Mayo, K. E., Warren, R., and Palmiter, R. D. (1982) The mouse metallothionein-I gene is transcriptionally regulated by cadmium following transfection into human or mouse cells. *Cells* **29,** 99–108.
6. Hofer, E., Hofer-Warbinek, R., and Darnell, J. E., Jr. (1982) Globin RNA transcription: a possible termination site and demonstration of transcription control correlated with altered chromatin structure. *Cell* **29,** 887–893.
7. Gurney, T., Jr., Sorenson, D. K., Gurney, E. G., and Wills, N. M. (1982) SV40 RNA: filter hybridization for rapid isola-

tion and characterization of rare RNAs. *Anal. Biochem.* **125,** 80–90.

8. Burton, K. (1969) Spectral data and pK values for purines, pyrimidines, nucleosides, and nucleotides. In *Data for Biochemical Research* (Dawson, R. M. C., Elliott, D. C., Elliott, W. H., and Jones, K. M., eds.) 2nd ed. Oxford University Press, Oxford and New York.

# Chapter 24

# Synthesis of Double-Stranded Complementary DNA from Poly(A)$^+$mRNA

**_R. McGookin_**

*Inveresk Research International Limited, Musselburgh, Scotland*

## Introduction

The use of avian myeloblastosis virus reverse transcriptase (AMV RTase) to produce DNA copies of mRNA templates is a common and well-documented method (*1–3*). Briefly, the method involves synthesis of a complementary DNA strand to the mRNA from a short double-stranded region, usually provided by using an oligo(dT) primer on poly(A)$^+$RNA. The enzyme does not always produce full length transcripts, but all the complementary strands are finished off with a short hairpin loop. This provides a ready-made primer for second strand synthesis, useful whether this is to be performed by more reverse transcriptase or by *E. coli* DNA polymerase 1 (pol 1). An idealized picture is shown in Fig. 1. Before the double-stranded cDNA (ds cDNA) copy can be cloned it is neces-

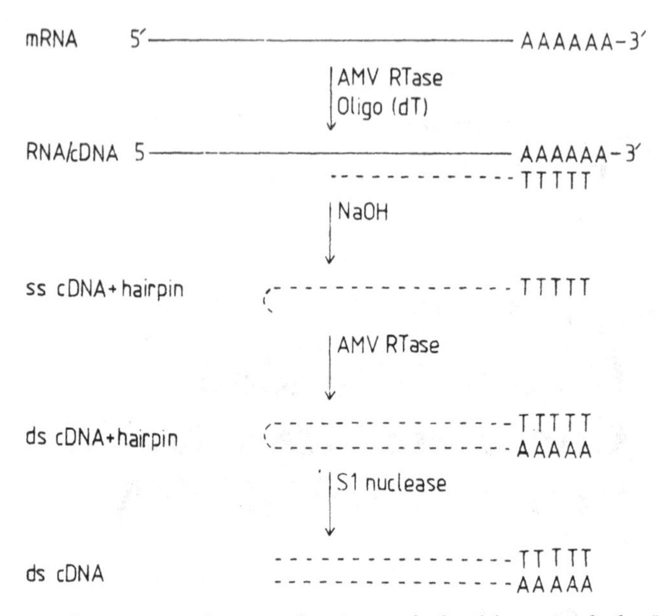


Fig. 1. Stages in the production of double-stranded cDNA from poly(A)$^+$mRNA. The original RNA is represented by a solid line, while the cDNA is represented by a dashed line. Note that this diagram is not intended as an accurate representation of the enzymatic processes involved, but as a general guide to the principles of cDNA synthesis.

sary to remove this hairpin loop using the single-strand specific nuclease S1.

The method detailed below uses AMV RTase for second strand synthesis. Although this generally results in a population of shorter ds cDNAs, the yield is higher per microgram of ss cDNA inputed. $^{32}$P-label is used in the first strand synthesis reaction and the second strand synthesis is estimated from the S1 resistance data Alternatively, a $^{3}$H-label can be used to measure second strand synthesis. Details of practical procedures are divided into three sections: First Strand Synthesis; Second Strand Synthesis; and S1 Nuclease Treatment.

## Materials

All the solutions and reagents mentioned below should be prepared from sterile stocks and, where possible, should themselves be autoclaved.

## First and Second Strand Synthesis

1. 10 × TMKD (0.5*M* Tris-Cl, pH 8.3, 80 m*M* $MgCl_2$, 4 m*M* DTT, 0.4*M* KCl): The stock buffer is prepared using double distilled or distilled deionized water (dd-$H_2O$) and stored in 1 mL aliquots at −20°C.
2. 10 × dNTPs: 10 m*M* aqueous solutions of each of the four bases in DNA stored at −20°C.
3. Oligo$(dT)_{12-18}$ (0.1 mg/mL): Stored at −20°C.
4. Poly$(A)^+$RNA (0.1 mg/mL): The material should have been purified by at least two passages over oligo(dT)-cellulose. Stored at −70°C.
5. AMV RTase (2 U/μL): The best material comes from Life Sciences, Gulfport, Florida, and is diluted to this strength with 0.2*M* K phosphate, 50% glycerol, 2 m*M* DTT, 0.2% Triton X-100. This diluted material together with the original stock is stored at −20°C.
6. $^{32}$P-dNTP: $^{32}$P-labeled nucleotide at the highest available specific activity.
7. 0.25*M* EDTA, pH 7.4.
8. 10% (w/v) SDS.
9. 1*M* NaOH.
10. 1*M* acetic acid.
11. Tris Buffered Saline (TBS): 150 m*M* NaCl, 50 m*M* Tris-Cl, pH 7.5.
12. TCA reagent: Equal volumes of 100% (w/v) TCA, saturated sodium pyrophosphate, and saturated $NaH_2PO_4$ are mixed to give a stock reagent that is stored at 4°C.
13. Phenol reagent (50% redistilled phenol/48% chloroform/2% isoamyl alcohol saturated with TBS) is stored at 4°C protected from light.
14. 10 × Column Buffer (1*M* sodium acetate, pH 7.5). Beware when adjusting the pH of this solution since acetate is a poor buffer at pH 7.5 and it is easy to overshoot. Store as sterile stock at 4°C.

## S1 Nuclease Treatment

1. 10 × S1 Buffer: (0.3*M* sodium acetate, pH 4.5; 0.5*M* NaCl, 10 m*M* $ZnSO_4$, 50% (v/v) glycerol): This solution is stored in 1 mL aliquots at −20°C.

2. S1 Nuclease (4 U/μL stored in 20 m*M* Tris-Cl, pH 7.5; 50 m*M* NaCl; 0.1 m*M* $ZnCl_2$; 50% glycerol): This enzyme is stable at −20°C.

# Method

## *First Strand Synthesis*

1. First strand synthesis is carried out in a final volume of 50 μL in a sterile, siliconized plastic centrifuge tube. Siliconized plastic and glassware should be used throughout this synthesis since ss cDNA is particularly susceptible to nonspecific absorption. Fifty microCuries of $^{32}P$-label is dried into the reaction tube under vacuum. The components of the reaction mix are then added as shown in Table 1. The reaction is carried out for 1 h at 45°C.
2. The reaction is stopped on ice and an aliquot taken for TCA precipitation (*see* Item 3 below). To the remaining reaction mix is added 4.5 μL of 0.25*M* EDTA, 1.1 μL of 10% SDS (to give 20 mM and 0.2%, respectively), and 7.5 μL of 1*M* NaOH and the tube is then placed at 37°C overnight to hydrolyze the RNA.
3. The 1 or 2 μL aliquot used for measuring incorporation is added to 200 μl of 1 m*M* EDTA on ice using a

Table 1
Reaction Mix for First Strand Synthesis

| Stock[a] | Vol, μL | Final concentration |
|---|---|---|
| 10 × TMKD | 5 | 50 m*M* Tris-Cl, pH 8.3; 8 m*M* $MgCl_2$; 0.4 m*M* DTT; 40 m*M* KCl |
| 10 m*M* dNTPs | 5 μL of each | 1 m*M* |
| 0.1 mg/mL oligo$(dT)_{12-18}$ | 2.5 | 5 μg/mL |
| dd-$H_2O$ | 6.5 | |
| 0.1 mg/mL poly$(A)^+$ RNA | 10 | 20 μg/mL |
| 2 U/μL AMV RTase | 6 | 240 U/mL |

[a]All stock solutions are kept on ice except the AMV RTase, which remains at −20°C. The enzyme is aliquoted last straight from the freezer. The reaction mix is kept on ice until addition of the enzyme, after which it is thoroughly mixed and placed in a water bath at 45°C.

microsyringe. To this is added 200 μL of TCA reagent and the tube is left on ice for at least 10 min. The precipitate is collected on Whatman GF/A or GF/C discs in a Millipore filtration apparatus under gentle vacuum. The tube is rinsed three times with 5% (w/v) TCA (a wash bottle is useful for this) and finally the filter is rinsed with more 5% TCA. The filter is dried and counted using a toluene-based scintillant.

4. After hydrolysis of the RNA, the alkali is neutralized with 7.5 μL of 1*M* acetic acid and 50 μL of TBS. A 100 μL volume of phenol reagent is added, the tube wrapped in parafilm, and the contents vortexed thoroughly for at least 30 s. The emulsion is separated by a 2 min spin in a high speed microfuge (12,000*g*), the aqueous (upper) phase is removed, and the organic layer is reextracted with 50 μL of TBS. The aqueous phases are combined.
5. A Sephadex G50 (Fine) column is prepared in a disposable 10 mL pipet and washed through with 10–20 μg of sheared DNA or poly(A)$^-$ RNA to fill any nonspecific binding sites. The column buffer is 0.1*M* sodium acetate, pH 7.5. The aqueous phase from the phenol extraction is loaded on the column and 5-drop fractions are collected. These samples are checked by counting without scintillant using a $^3$H-channel (Cerenkov counting), the excluded fractions are pooled, 2 μL of a 2 mg/mL tRNA carrier is added, and the ss cDNA plus tRNA is precipitated with 2.5 vol ethanol at −20°C overnight.

## Second Strand Synthesis

1. The ss cDNA plus carrier is spun down at 12,000*g* for 10 min and the pellet washed once with 500 μL of 80% ethanol before drying *in vacuo*. The pellet is dissolved in 13 μL of dd-$H_2O$ and the second-strand reaction mix made up as described in Table 2. The reaction is carried out at 45°C for 1 h.
2. The reaction is stopped on ice, and EDTA and SDS are added as in first strand synthesis. The mix is then extracted with phenol and chromatographed as before. Finally, the pooled excluded fractions are precipitated by ethanol at −20°C overnight.

Table 2
Reaction Mix for Second Strand Synthesis[a]

| Stock[a] | Vol, µL | Final concentration |
|---|---|---|
| 10 × TMKD | 5 | 50 m*M* Tris-Cl, pH 8.3; 8 m*M* $MgCl_2$; 0.4 m*M* DTT; 40 m*M* KCl |
| 10 m*M* dNTPs | 5 µL of each | 1 m*M* |
| ss cDNA | 13 | Various |
| 2 U/mL AMV RTase | 12 | 480 U/mL |

[a]The same conditions apply as for first strand synthesis. The reaction is carried out in the tube used to precipitate the ss cDNA that is first dissolved in 13 µL of dd-$H_2O$. The AMV RTase is added to start the reaction and is aliquoted directly from the freezer.

## *Nuclease Treatment*

1. The ds cDNA prepared in the previous section is spun down at 12,000*g* for 10 min and washed twice with 250 µL of 80% ethanol. The dried pellet is dissolved in 90 µL of dd-$H_2O$ before 10 µL of 10 × S1 buffer plus 1 µL of S1 nuclease (4 U/µL) is added. The reaction proceeds at 45°C for 40 min after removing an aliquot for TCA precipitation.
2. At the end of the incubation, another sample is taken and the reaction is stopped with 9 µL of 0.25*M* EDTA and 2.2 µL of 10% SDS. A 3*M* NaCl solution is added to 0.3*M* and 2.5 vol of ethanol. The ds cDNA is stored at −20°C.

## Notes

1. There are several steps in the above procedure where it is possible, if desired, to speed up the process. The hydrolysis of the RNA with alkali after first strand synthesis may be carried out at 68°C for 30 min instead of at 37°C overnight. The ethanol precipitation steps can be carried out in a dry ice/ethanol bath for 2 h.
2. It is important to ensure that the substrate is clean for each of the above reactions, hence the plethora of desalting columns, ethanol precipitations, and washes. The column used for these operations must be of an ad-

equate size to ensure complete separation of small molecules from the cDNA. The S1 nuclease is particularly sensitive to inhibition by deoxynucleotides giving an erroneous estimate of ds cDNA and failure to provide a suitable substrate for further cloning operations.

3. If problems are encountered with ss cDNA sticking to the Sephadex or other column components, it may help to extract the reaction with phenol immediately after first strand synthesis without hydrolyzing the RNA. This leaves a cDNA–RNA double strand hybrid to pass over the column. The RNA is then hydrolyzed before ethanol precipitation. One of the conveniences of using $^{32}P$ as a first-strand label is being able to follow the progress of the cDNA with a hand-held radiation monitor.
4. An improved size of transcript in the first strand synthesis can be obtained by adding a few units of human placental RNase inhibitor (RNasin) (*4*). A series of test reactions should be set up to determine the optimum ratio of AMV RTase to inhibitor, the results being determined from a dissociating gel system. This inhibitor may work by reducing the ribonuclease H associated with AMV RTase (*5*).
5. Another use for reverse transcriptase is to produce radioactive probes for hybridization studies, such as Southern transfers (*6*) (and *see* Chapter 4). The first strand synthesis is performed exactly as described above, although more label may be used. The RNA is hydrolyzed off and the labeled ss cDNA can be used to detect complementary sequences on filters.

# References


1. Kacian, D. L., and Myers, J. C. (1976) Synthesis of extensive, possibly complete, DNA copies of poliovirus RNA in high yields at high specific activities. *Proc. Natl. Acad. Sci. USA* **73,** 2191–2195.
2. Buell, G. N., Wickens, M. P., Payvar, F., and Schimke, R. T. (1978) Synthesis of full length cDNAs from four partially purified oviduct mRNAs. *J. Biol. Chem.* **253,** 2471–2482.
3. Okayama, H., and Berg, P. (1982) High-efficiency cloning of full length cDNA. *Mol. Cell. Biol.* **2,** 161–170.

4. Blackburn, P., Wilson, G., and Moore, S. (1977) Ribonuclease inhibitor from human placenta. Purification and preparation. *J. Biol. Chem.* **252,** 5094–5910.
5. Berger, S. L., Wallace, D. M., Puskas, R. S., and Eschenfeldt, W. H. (1983) Reverse transcriptase and its associated ribonuclease H: Interplay of two enzyme activities controls the yield of single stranded cDNA. *Biochemistry,* **22,** 2365–2373.
6. Southern, E. M. (1975) Detection of specific sequences among DNA fragments separated by gel electrophoresis. *J. Mol. Biol.* **98,** 503–517.

# Chapter 25

# Plasmid DNA Isolation by the Cleared Lysate Method

***Stephen A. Boffey***

*Division of Biological and Environmental Sciences, The Hatfield Polytechnic, Hatfield, Hertfordshire, England*

## Introduction

The cleared lysate method of plasmid isolation is commonly used to extract relatively small plasmids (up to about 20 kb) from gram-negative bacteria such as *Escherichia coli*. It relies on a very gentle lysis of the bacteria to release small molecules, including very compact, supercoiled plasmid, into solution, while trapping larger molecules, such as chromosomal DNA fragments, in the remains of the cells. A high-speed centrifugation pellets cell debris and trapped chromosomal DNA, to produce a 'cleared lysate' highly enriched for plasmid.

The yields of plasmids such as pBR322, which contain a Co1E1 replicon, can be increased by amplification, using chloramphenicol to inhibit replication of chromosomal DNA, but not of plasmids. Such amplification can result in up to 3000 copies of plasmid per cell. After lysis, clearing of the lysate, deproteinization, CsCl density gradient ultracentrifugation, and dialysis, up to 1 mg of supercoiled plasmid can be obtained from 1 L of bacteria, in a form suitable for restriction or transformation.

The procedure is developed from that of Clewell and Helinski (*1*), and assumes the presence in the plasmid of a gene for ampicillin resistance.

## Materials

1. LB Broth. Yeast extract 5 g, NaCl 10 g, tryptone 10 g, distilled water 1 L. After autoclaving add 1 mL of 20% glucose (filter sterilized) to each 100 mL of broth.
2. Ampicillin. Prepare stock of 50 mg/mL in sterile distilled water, using a little NaOH to dissolve the ampicillin initially. Add 1 mL of this stock to each liter of LB broth after the broth has been autoclaved, to give a final concentration of 50 μg/mL.
3. Chloramphenicol. Prepare stock of 150 mg/mL in absolute ethanol. Add 1 mL to each liter of LB broth in step '2' of the method section, to give 150 μg/mL.
4. TES. Tris base 30 m*M*, $Na_2EDTA$ 5 m*M*, NaCl 50 m*M*, pH 8.0.
5. Tris/sucrose. Tris 50 m*M*, sucrose 25% (w/v), pH 8.0.
6. 5 mg Lysozyme/mL, in Tris 0.25*M*, pH 8.0. Prepare immediately before use.
7. EDTA 0.25*M*, pH 8.0.
8. Sodium dodecyl sulfate (SDS) 10% (w/v).
9. Tris 0.25*M*, pH 8.0.
10. Tris base 2.0*M*, no pH adjustment.
11. Phenol (redistilled or really fresh AR grade) mixed with chloroform, 1:1 (v/v). Store indefinitely in dark bottles at 4°C.
12. Potassium acetate, 4.5*M*.
13. Absolute ethanol, stored at −20°C.
14. SSC. NaCl, 0.15*M*; $Na_3$citrate, 0.015*M*. Usually used at tenfold dilution (0.1 × SSC).
15. Ethidium bromide 5 mg/mL. Use gloves when handling this powerful mutagen.
16. Isoamyl alcohol.

17. All glassware, centrifuge tubes, syringes, and so on, which will come into contact with the DNA, should be autoclaved to destroy any DNase activity. The centrifugations in steps 13 and 25 will require an ultracentrifuge capable of ~40,000 rpm. For centrifugation of CsCl in step 25 it is advisable to use a titanium alloy rotor, since this will not be harmed by accidental contact with CsCl. It is not unknown for centrifuge manufacturers to incorporate a UV-absorbing compound in their polycarbonate centrifuge tubes. This certainly prolongs the lives of tubes if they are exposed to a lot of UV radiation, but it makes them useless for CsCl ultracentrifugation of DNA, where bands of DNA plus bound dye are to be revealed by their fluorescence. Always specify 'UV-transparent' tubes.

# Method

1. Inoculate 100 mL of LB medium containing 50 μg ampicillin/mL with a loop of bacteria and incubate overnight at 37°C.
2. The next morning inoculate 1 L of prewarmed LB/ampicillin with 40 mL of the overnight culture, and incubate with shaking at 37°C. After about 1.5 h remove a few milliliters of suspension and measure its absorbance at 660 nm, using LB as blank. Repeat at intervals until $A_{660}$ is about 0.4 (be prepared for a rapid increase in absorbance; doubling time is about 30 min), then add 1.0 mL of a 15% (w/v) solution of chloramphenicol in ethanol, giving a final concentration of 150 μg/mL.
3. Incubate this culture, with shaking, at 37°C overnight (for at least 16 h; longer will do no harm).
4. Harvest the cells by centrifugation at 2500*g* in six 250 mL bottles, at 4°C, for 10 min. Note that all relative centrifugal forces are given as '$g_{average}$'.
5. Pour off the supernatants, which should be autoclaved before disposal. Resuspend the pellets in 30 mL chilled TES buffer per bottle. This is most easily achieved by violent shaking of each (sealed) bottle,

using a vortex mixer, until the pellet is resuspended in its own 'juices' to give a smooth paste, and then the TES is added to give a homogeneous suspension, free of cell clumps.

6. Transfer the suspensions into six 50 mL centrifuge tubes, and centrifuge at 3000*g* for 10 min at 4°C.
7. Decant supernatants and resuspend each pellet in 2.5 mL of chilled Tris/sucrose.
8. Transfer suspensions to a chilled conical flask (giving 15 mL total volume).
9. Add 3 mL of lysozyme solution, then 5 mL of EDTA (0.25*M*, pH 8). Swirl on ice for 10 min to allow the EDTA and lysozyme to weaken cell walls.
10. Add 6 mL of 10% (w/v) SDS and IMMEDIATELY give the flask a single swirl to ensure mixing. Treat the suspension gently from now on, since it is important not to damage high molecular weight DNA released from the cells.
11. Incubate at 37°C (definitely no shaking) until the suspension loses its cloudy appearance as a result of cell lysis. This often occurs after a minute or two, but may take over 30 min. A good way to test for cell lysis is to hold the flask at eye level, tilt it gently, and watch for a highly viscous 'tail' sliding down the glass behind the bulk of the suspension. This high viscosity results from the release of high molecular weight DNA from the bacteria following lysis by SDS.
12. Tip this lysate gently into two 35 mL, thick walled, polycarbonate centrifuge tubes. Because of its gel-like consistency, the lysate will probably have to be split in two by *cutting* with a pair of flamed scissors.
13. Use Tris (0.25*M*, pH 8) to balance the tubes, and then centrifuge at 120,000*g* for at least 1 h at 20°C. This will 'clear the lysate' by forming a translucent pellet containing cell debris and, tangled with the debris, much of the high molecular weight chromosomal DNA. Consequently the supernatant will contain most of the plasmid and relatively little chromosomal material; it will also contain RNA and proteins.
14. Carefully decant the supernatants into a measuring cylinder, and note the total volume.

15. Transfer the pooled supernatants into a conical flask and add 0.1 times the volume of 2.0*M* Tris base (pH not adjusted). Then double the volume by adding phenol/chloroform (1:1, v/v).
16. Shake this mixture sufficiently to form an emulsion, and keep it emulsified by occasional shaking for 10 min.
17. Centrifuge the emulsion in glass centrifuge tubes at top speed in a bench centrifuge (about 3500*g*) for 10 min to separate the aqueous and organic phases. Transfer the upper (aqueous) layers into a clean conical flask, using a Pasteur pipet; be careful not to transfer any of the white precipitate at the interface (this contains denatured protein).
18. Add an equal volume of phenol/chloroform to the aqueous solution, and re-extract as in steps 16 and 17. This should remove almost all protein from the nucleic acids solution.
19. Measure the volume of aqueous phase finally collected, and add potassium acetate to give a 0.9*M* solution (add 0.25 vol of 4.5*M* potassium acetate). This is needed to ensure quantitative precipitation of low concentrations of DNA in the next step.
20. Work out the new volume of solution and add 2 vol of chilled ethanol. Mix thoroughly, then transfer to 50 mL polypropylene centrifuge tubes. Leave at −20°C for at least an hour to allow precipitation of DNA and some RNA.
21. Centrifuge at 12,000*g* for 10 min at 0°C. Decant the supernatants and, keeping the tubes inverted, blot off any remaining drops of ethanol. Remove traces of ethanol by evaporation in a vacuum desiccator.
22. Dissolve the precipitates (don't worry if they seem invisible) in a total volume of 14 mL of 0.1 × SSC by gently swirling the liquid round the tubes. Do *not* use a vortex mixer.
23. Transfer the solution into a flask containing 15.4 g CsCl, and swirl until this is completely dissolved.
24. Add 1.6 mL of ethidium bromide (5 mg/mL).
25. This solution should now be centrifuged for at least 40 h at 140,000*g* in a swing-out rotor, at 20°C. If your ro-

tor will not hold 20 mL total volume, the volumes of 0.1 × SSC and ethidium bromide, and weight of CsCl (steps 22–24) can be reduced in proportion; however, if this is taken too far the CsCl gradient will be overloaded. Avoid rotors with long, narrow tubes; short, wide tubes give the sharpest bands.

26. After centrifugation view the tubes using long wavelength UV light. Two well-defined bands should be seen near the middle of the tube, separated by about 1 cm. The lower band is supercoiled, covalently closed, circular plasmid. If the cleared lysate procedure has worked well, this band should be more intense than the upper one, which contains fragments of chromosomal DNA and also linear and open circle forms of plasmid.
27. Although it is possible to recover the plasmid band by side-puncturing the centrifuge tube, the easiest way is to draw off material from above. First draw off the upper part of the gradient, including the upper DNA band, using a Pasteur pipet. Then fix the centrifuge tube beneath a syringe fitted with a wide bore needle, and lower the syringe (or raise the tube) slowly until the needle tip is just below the plasmid band. Provided both syringe and tube are both firmly fixed, you should have no difficulty in sucking the plasmid band into the syringe. When all the plasmid has been removed a very thin band will remain in the tube; this is an optical effect caused by the sudden change in refractive index where a 'slice' of continuous gradient has been removed. Do not try to collect this 'band'!
28. Transfer the plasmid material into a 1.5 mL polypropylene tube (approx. 0.5 mL/tube), and add almost enough isoamyl alcohol to fill each tube. Cap the tubes and invert them several times. The alcohol will become pink as ethidium bromide partitions into it. Remove the upper (alcohol) layer, and re-extract with fresh alcohol. Repeat this until no color can be detected in the alcohol, then once more to be sure.
29. To remove CsCl from the plasmid solution, transfer it into dialysis tubing (pretreated by boiling in 10 m*M* EDTA for 15 min, followed by two 15 min treatments in boiling distilled water) and dialyze against three changes of 500 mL 0.1 × SSC over about 16 h at 4°C.

Wear disposable gloves when handling dialysis tubing.

30. The plasmid is now ready for use. If a more concentrated preparation is needed, concentrate it by ethanol precipitation, as described in steps 19–22. If the plasmid must be in a buffer other than 0.1 × SSC, use that buffer for dialysis.

# Notes

1. This method usually produces a high yield of supercoiled plasmid, free of chromosomal DNA. If, however, the separation of bands after CsCl ultracentrifugation is not considered satisfactory (e.g., if overloading has caused broadening of bands), material recovered from the lower band can be recentrifuged after addition to fresh CsCl/0.1 × SSC/ethidium bromide prepared as in steps 22–24. Because there will be no problems caused by precipitates in this second centrifugation, it can be carried out in an angle rotor for only 16 h at 140,000*g*, 20°C.
2. If yields of plasmids are low the cause is most likely to be poor lysis of the cells. This can usually be rectified by prolonging step 11 and/or by freezing and thawing between steps 7 and 8; however, remember that you are aiming at *incomplete* lysis in order to pellet most of the chromosomal DNA during clearing of the lysate.
3. Owing to the long ultracentrifugation this procedure takes about 2.5 d if only one CsCl run is included, or 3.5 d with a second spin. However, the product is very pure, and can be stored frozen in 0.1 × SSC for several months. It is best to freeze the plasmid in small aliquots, as repeated cycles of freezing and thawing will damage the DNA.

# References

1. Clewell, D. B., and Helinski, D. R. (1971) Properties of a supercoiled deoxyribonucleic acid-protein relaxation complex and strand specificity of the relaxation event. *Biochemistry* **9**, 4428–4440.

# Chapter 26

# Plasmid DNA Isolation (Sheared Lysate Method)

***J. W. Dale and P. J. Greenaway***

*Department of Microbiology, University of Surrey, Guildford, Surrey and Molecular Genetics Laboratory, PHLS Centre for Applied Microbiology and Research, Porton Down, Salisbury, Wilts., United Kingdom*

## Introduction

The cleared lysate method (*see* Chapter 25) is not usually very effective for isolation of plasmids larger than about 20 kb. Recovery of plasmid DNA is often poor, presumably because high molecular weight plasmids are removed by the clearing spin. An alternative procedure, therefore, is to load the complete cell lysate onto a cesium chloride–ethidium bromide gradient that will separate the plasmid DNA from the chromosomal material and also from other cell components.

However, the cell lysate is extremely viscous and in order to get good bands on the gradient the viscosity of the lysate must be reduced. This can be done by repeated passage through a syringe needle, which shears the chromosomal DNA. The supercoiled state of the plasmid renders it less susceptible to shearing. This method has been used successfully with plasmids up to 60 kb; larger plasmids become too susceptible to shearing for this approach to be effective. The sheared lysate method is also useful for the isolation of plasmids from bacteria other than standard laboratory strains of *Escherichia coli*, e.g., from environmental isolates, since the lysis conditions are somewhat more robust than those used for the cleared lysate procedure.

This method is based on procedures originally described by Barth and Grinter (*1*) and Bazaral and Helinski (*2*).

# Materials

1. TES buffer: 0.05*M* Tris-HCl, 0.005*M* EDTA, 0.05*M* NaCl, pH 8.0. Autoclave and store at 4°C for maximum shelf life.
2. Spheroplast mix (make fresh just before use): Add 5 mg of ribonuclease to 10 mL of TES. Heat at 80°C for 15 min. Add 1.0 g of sucrose (while hot) and dissolve. Allow to cool to room temperature and add 10 mg of lysozyme.
3. Sarkosyl: 2% solution of sarkosyl in water. This is stable at room temperature.
4. Ethidium bromide: 20 mg/mL in distilled water.

**CAUTION**: Ethidium bromide is mutagenic and a potential carcinogen. Wear gloves when handling it and be especially careful when weighing out the dry powder.

# Method

1. Set up an overnight starter culture in L broth, plus an appropriate selective antibiotic, if applicable.
2. Inoculate 300 mL of L broth (plus antibiotic) in a 1 L flask and grow, with shaking, to an $A_{600}$ of less than 0.8. This corresponds to a cell density of about $5 \times 10^8$/mL. This should take about 2–3 h.
3. Cool the flask on ice and recover the cells by centrifugation (15,000*g* for 10 min.) Wash the pellet with 50 mL of TES.
4. Resuspend the pellet in 10 mL of spheroplast mix. Transfer to a 100 mL conical flask and incubate in a 37°C water bath for 10 min.
5. Chill in an ice bath for 5 min. Add 5 mL of sarkosyl and mix by gentle pipetting. The suspension should now be very viscous, but may remain turbid.
6. Add 10 mL of TES at room temperature (this may be omitted if the original cell density was low). Shear by passage through a 19G syringe needle 20 times. The solution should become markedly less viscous (*see* Notes below).
7. Measure the volume and add solid CsCl at the rate of 0.95 g/mL. Mix gently to dissolve. Add 0.5 mL of ethidium bromide, mix and distribute into ultracentrifuge tubes.
8. Centrifuge at 100,000*g* (e.g., 40,000 rpm in a Beckman Ti50 or Ti75 rotor) for at least 36 h at 18°C.
9. Remove the tubes from the rotor and examine them with long wave ultraviolet light. Two DNA bands should be visible; the lower (plasmid) band should be a thin sharp band, while the upper (chromosomal) band will be broader and fuzzy. At the top of the tube there will be a red pellicle formed from denatured protein complexed with the ethidium bromide.
10. The DNA bands are removed with a syringe needle, either through the side of the tube or from the top. It is advisable to remove most of the chromosomal band before attempting to take the plasmid DNA. The fibrous nature of the chromosomal band means that it

can be very easily sucked into the syringe needle even if the tip is not within the visible band. Alternatively, puncture the bottom of the tube and collect the drops corresponding to the plasmid DNA.

11. The DNA is then further purified as described in Chapter 25, using isopropanol to extract the ethidium bromide, dialysis to remove the cesium chloride, finishing with ethanol precipitation.

## Notes

1. The main problem with this procedure is the shearing step. Too little shearing results in a viscous mixture that will not separate properly on the gradient; too much is likely to result in loss of plasmid DNA. There is no easy way of knowing the extent of shearing required; it has to be learned by experience. The reduction in viscosity is revealed by the decrease in effort needed to pass the solution through the syringe needle. If the plasmid is known to be large (say 50 kb or more), then it is advisable to reduce the amount of shearing (for example, by using a 10 mL pipet instead of a syringe needle), and accept that the separation on the gradient will not be complete. For extremely large plasmids (100 kb or more), which are likely to be unstable, then alternative methods based on, e.g., alkaline sucrose gradients (3) are advisable.
2. If plasmid DNA of high purity is required, it is often necessary to use a second CsCl gradient. Immediately after removing the DNA band from the first gradient, transfer it to a fresh ultracentrifuge tube. Fill the tube with more CsCl solution of the correct density, add ethidium bromide and recentrifuge as before.

## References

1. Barth, P. T. and Grinter, N. J. (1974) Comparison of the deoxyribonucleic acid molecular weights and homologies of plasmids conferring linked resistance to streptomycin and sulphonamides. *J. Bacteriol.* **120,** 618–630.

2. Bazaral, M., and Helinski, D. R. (1968) Circular DNA forms of colicinogenic factors E1, E2 and E3 from *Escherichia coli*. *J. Mol. Biol.* **36,** 185–194.
3. Wheatcroft, R., and Williams, P. A. (1981) Rapid methods for the study of both stable and unstable plasmids in Pseudomonas. *J. Gen. Microbiol.* **124,** 433–437.

# Chapter 27

# Small-Scale Plasmid DNA Preparation

***J. W. Dale and P. J. Greenaway***

*Department of Microbiology, University of Surrey, Guildford, Surrey and Molecular Genetics Laboratory, PHLS Centre for Applied Microbiology and Research, Porton Down, Salisbury, Wilts., United Kingdom*

## Introduction

For the initial characterization of a recombinant plasmid, it is necessary to determine the size of the plasmid or, preferably, the size and characteristics of the insert itself. A method is therefore required for the simultaneous preparation, from a number of isolates, of plasmid DNA in a state sufficiently pure for restriction enzyme digestion. The requirements of such a procedure are:

(i) A simple method for rapid lysis of the bacterial cells.
(ii) Separation of plasmid from chromosomal DNA.

(iii) Removal of proteins and of other components of the cells that might interfere with restriction enzyme treatment.
(iv) Removal of detergents, salts, etc. used in the process.

The procedure outlined below is based on that published by Birnboim and Doly (*1*). This involves treating the cells with a lysozyme–EDTA mixture to weaken the cell walls; lysis is completed by the addition of alkaline sodium dodecyl sulfate (SDS). Chromosomal DNA will be extracted in the form of linear fragments; the high pH weakens the hydrogen bonds holding the two chains together, which therefore separate. On rapid neutralization, this denatured DNA forms an insoluble network which can therefore be removed by centrifugation. In contrast, intact plasmids, which are supercoiled covalently closed circular (CCC) molecules, behave differently. The two strands are unable to separate fully even with all the hydrogen bonds disrupted. Denatured plasmid molecules will therefore rapidly renature when the pH is lowered and will remain in solution. The high salt concentration also results in the precipitation of SDS–protein complexes. Most of the cell proteins, together with much of the SDS added to lyse the cells, can therefore be removed by centrifugation. Plasmid DNA is further purified by the subsequent ethanol precipitation step.

## Materials

All buffers and other reagents that are to be stored should be autoclaved, used aseptically, and stored at 4°C. An additional precaution is to pass the solution through a Millipore filter, which will ensure freedom from dust particles etc.

1. Lysis solution: 25 m*M* Tris-HCl, pH 8.0; 10 m*M* EDTA, pH8.0; 50 m*M* glucose. Add 2 mg/mL lysozyme just before use.
2. Alkaline SDS: 1% (w/v) SDS in 0.2*M* sodium hydroxide. The SDS will precipitate if stored in the refrigera-

tor, but can be redissolved by heating to 65°C in a waterbath.

3. High salt buffer: 3*M* sodium acetate, adjusted to pH 4.8 with glacial acetic acid.
4. Low salt buffer: 0.1*M* ammonium acetate, pH 6.0.
5. TE buffer: 0.01*M* Tris-HCl, pH 8.0; 1 m*M* EDTA.

# Method

1. Inoculate about 2 mL of broth with the colony to be tested and incubate overnight at 37°C with shaking. Antibiotics can be added (if appropriate) to ensure plasmid retention or amplification.
2. Transfer 1 mL of the culture to a large microfuge tube and pellet the cells. Resuspend the pellet in 100 μL of lysis solution and store it on ice for 30 min. If the pellet is difficult to resuspend initially, remix after about 10 min.
3. Add 200 μL of alkaline SDS (at room temperature) and keep the mixture on ice for 5 min. The suspension should first become clear and slightly viscous, but may then become cloudy because of the SDS precipitating as the suspension cools down.
4. Add 150 μL of 3*M* sodium acetate, mix gently, and store on ice for 60 min. Note that a heavy, coarse precipitate is formed.
5. Centrifuge for 5 min at room temperature. Transfer 400 μL of the supernatant to another tube, avoiding any contamination with the precipitate; if 400 μL cannot be withdrawn, settle for less. If there is insoluble material floating in the tube, it may be difficult to avoid contamination. In this case a second centrifugation is necessary.
6. Add 1 mL of ethanol to the supernatant and store at −70°C for at least 30 min. If a −70°C freezer is not available, use a dry ice–ethanol bath. Alternatively, a −20°C freezer can be used, but the time must be extended to at least 2 h, or (preferably) overnight.
7. Centrifuge for 5–10 min and carefully remove the supernatant using a drawn-out Pasteur pipet. Allowing the temperature to rise at this point is a common cause

of failure. The yield can be improved by refreezing the tubes, without taking off the ethanol, and then repeating the centrifugation. Alternatively, use a microcentrifuge in a coldroom. Resuspend the pellet (which should be barely visible) in 100 μL of 0.1*M* ammonium acetate, add 300 μL of cold ethanol and keep at −70°C for at least 30 min.

8. Centrifuge and discard the supernatant as before. Dry the pellet in a vacuum desiccator for about 15 min, resuspend in 100 μL of TE buffer and store at 4°C prior to analysis.

## Notes

1. Avoid the temptation to grow the cultures in the microfuge tube. There is not enough air space for a good yield to be obtained.
2. The number of isolates that can be handled at one time is usually determined by the capacity of your microcentrifuge; this usually means doing 12 clones (or multiples thereof) at a time.
3. Scaling up the process does not seem to work very well. If you only have a few isolates to test, but would like more DNA from each, it is better to put up replicates rather than increasing the volumes at each step.
4. A 10–20 μL volume of the final solution is usually sufficient to give a good picture on an agarose gel. There is usually some RNA present. If this is likely to obscure the insert band on the gel, treatment of the plasmid preparation with DNase-free ribonuclease is necessary.
5. Plasmid DNA prepared by this method is stable for a limited period of time only, i.e., not more than a few days.
6. The method is suitable for most commonly used plasmid vectors and *E. coli* host strains; we have also used it under other circumstances, e.g., with larger wild-type plasmids and environmental isolates, but it is not then quite so reliable. Other rapid methods are available which are more suitable for very large and/or unstable plasmids; see for example Wheatcroft and Williams (2).

7. The problems commonly encountered are:

   (a) Excessive contamination and/or failure of the restriction digest. This is usually because of contamination by the precipitate at step 5.
   (b) The failure of the restriction digest can also be caused by carryover of salt, owing to failure to remove all of the ethanol following precipitation of the DNA. Reprecipitate the DNA and try again!
   (c) A substantial insoluble precipitate is formed after ethanol precipitation. This is usually also due to contamination by the precipitate at step 5, and can be partially resolved by adding TE buffer to dissolve the DNA, centrifuging, and then using the clear supernatant.
   (d) No plasmid DNA obtained. This can result from:
   Absence of plasmid or an unstable plasmid (use an antibiotic in the growth medium).
   Failure of ethanol precipitation (the conditions outlined must be followed carefully).
   Nuclease contamination (wear gloves; check buffers and reagents for contamination).

8. It is quite feasible to carry the procedure through, perform restriction digests and run an agarose gel all on the same day. When trying the procedure for the first time it is advisable to leave one of the ethanol precipitation stages overnight and finish the preparation the next day.
9. Note that this procedure can also be used for screening the plasmid-like replicative forms of phage vectors based on M13 and similar phages.

# References


1. Birnboim, H. C., and Doly, J. (1979) A rapid alkaline extraction procedure for screening recombinant plasmid DNA. *Nucl. Acids Res.* **7,** 1513–1523.
2. Wheatcroft, R., and Williams, P. A. (1981) Rapid methods for the study of both stable and unstable plasmids in Pseudomonas. *J. Gen. Microbiol.* **124,** 433–437.

# Chapter 28

# Preparation of Chromosomal DNA from *E. coli*

***J. W. Dale and P. J. Greenaway***

*Department of Microbiology, University of Surrey, Guildford, Surrey and Molecular Genetics Laboratory, PHLS Centre for Applied Microbiology and Research, Porton Down, Salisbury, Wilts., United Kingdom*

## Introduction

This chapter describes a simple and rapid way of extracting and purifying chromosomal DNA from *E. coli* and many other species of bacteria. This procedure is essentially a simplified version of that described by Marmur in 1961 (*1*). The cells are lysed by treatment with a detergent and the mixture is deproteinized by phenol–chloroform extraction. Further purification can be achieved by treatment with ribonuclease and proteinase K. The resulting DNA, free of protein and RNA contamination, is sufficiently pure to be used for restriction digestion and cloning, e.g., in the preparation of gene libraries.

## Materials

1. 20 × SSC buffer: 3*M* NaCl, 0.3*M* sodium citrate. Adjust pH to 7.0 with sodium hydroxide and sterilize by autoclaving. Store in aliquots at 4°C for maximum shelf life.
2. Double-strength SSC buffer (2 × SSC): Prepare just before use by a 1 in 10 dilution of 20 × SSC in water.
3. Phenol/chloroform: Phenol should ideally be redistilled under nitrogen before use. It should then be equilibrated with several changes of buffer (in this case, 20 × SSC) and stored at 4°C in a dark bottle. Discard any phenol showing a pink color. Oxidation of the phenol can be minimized (and its shelf life thereby increased), by the addition of 0.1% of 8-hydroxyquinoline. Prepare a 1:1 mixture (by volume) of phenol and chloroform immediately before use. A mixture of chloroform and isoamyl alcohol (24:1 v/v) can be used in place of chloroform.

   **CAUTION**: Phenol is highly corrosive and causes severe burns. Always wear gloves when using phenol. If any contact with the skin occurs, wash with soap and large volumes of water. Do NOT use ethanol.

4. Ribonuclease (DNase free).
5. Proteinase K.
6. Absolute ethanol.
7. TES buffer: 10 m*M* Tris-HCl, pH 8.0; 10 m*M* NaCl; 1 m*M* EDTA.

## Method

1. Inoculate 200 ml of L broth (in a 500 or 1000 mL conical flask) with a starter culture derived from a single colony. Incubate the culture overnight at 37°C with shaking.
2. Recover the cells by centrifugation (15,000*g* for 10 min). Resuspend the pellet in 20 mL of 20 × SSC buffer.

3. Add 200 mg of sodium dodecyl sulfate and rotate the suspension at room temperature until lysis has been achieved (overnight if necessary). The solution should become viscous.
4. Add an equal volume of phenol/chloroform to the resulting viscous suspension and mix gently but thoroughly. Denatured protein will collect at the interface.
5. Recover the aqueous phase, without contamination by the material at the interface and repeat the extraction until little protein is extracted (this requires at least three extractions). Note that if the precipitate is heavy, a lot of DNA will remain trapped in it. To obtain the maximum yield of DNA, back extract the organic phase from the first extraction with 20 × SSC and pool this with the aqueous phase for the subsequent extractions.
6. To the aqueous phase remaining after phenol/chloroform extraction, add two volumes of absolute ethanol to precipitate the DNA. The DNA can be collected by spooling onto a glass rod or by centrifugation.
7. Resuspend the DNA in 15 mL of 2 × SSC and add RNase to 50 μg/mL. Incubate at 37°C for 1 h.
8. Add proteinase K (to 50 μg/mL) and incubate at 37°C for a further hour.
9. Extract the resulting suspension with phenol and precipitate the DNA by adding 2.5 vol of absolute ethanol to the aqueous phase. Allow the DNA to precipitate overnight at −20°C.
10. Recover the DNA by centrifugation at 25,000*g* for 15 min at 4°C and carefully remove the ethanol. Wash the pellet with absolute ethanol at −20°C.
11. Dry the DNA in a vacuum desiccator and resuspend in 10 mL of TES. Measure the optical density of the solution at 280, 260, and 235 nm, and then store at 4°C.

## Notes

1. The principles of this procedure are applicable to most species of bacteria. However, the conditions of lysis

may need to be varied, e.g., for gram-positive bacteria such as *Staphylococci* it may be necessary to use lysozyme or another cell-wall degrading enzyme before SDS treatment. For other bacteria (e.g., *Mycobacteria*) heating the suspension to 65°C after adding SDS is often effective.

2. Another factor that must be taken into account is that many bacteria produce powerful nucleases that may not be inhibited by the detergent. Heating the suspension during the lysis procedure will usually help to destroy these nucleases; the addition of EDTA (by using a Tris-EDTA–sodium chloride buffer instead of SSC) also inhibits nuclease action. Note, however, that TES buffer has a lower ionic strength and sodium acetate must be added to a final concentration of 0.3*M* before ethanol precipitation of the DNA.
3. Traces of phenol or chloroform can be removed from the DNA before (or instead of) ethanol precipitation by extraction with water-saturated ether.
4. For isolation of relatively small fragments of DNA, the phenol extractions can be vortex mixed. If larger fragments are required, shearing must be avoided as much as possible. The organic and aqueous phases must be mixed gently, e.g., by rotation for several hours.

**CAUTION**: Ether is highly volatile and inflammable and should be used in an efficient fume cupboard. Any material containing even traces of ether should not be stored in a refrigerator unless it is internally spark-proofed. Ether and chloroform wastes must not be discarded down the sinks.

5. Good DNA solutions have $A_{260}:A_{280}$ and $A_{260}:A_{235}$ ratios greater than 1.7

## References

1. Marmur, J. (1961) A procedure for the isolation of deoxyribonucleic acid from micro-organisms. *J. Mol. Biol.* **3,** 208–218.

# Chapter 29

# Preparation and Assay of Phage Lambda


## *J. W. Dale and P. J. Greenaway*

*Department of Microbiology, University of Surrey, Guildford, Surrey and Molecular Genetics Laboratory, PHLS Centre for Applied Microbiology and Research, Porton Down, Salisbury, Wilts., United Kingdom*


## Introduction

Lambda, a temperate bacteriophage of *E. coli,* has two alternative modes of replication in sensitive cells, known as the lytic and lysogenic cycles. In the lytic cycle, after the lambda DNA enters the cells, various phage functions are expressed that result in the production of a large number of mature phage particles and cell lysis. In the lysogenic mode, which normally occurs in only a small proportion of the infected cells, the phage forms a more or less stable relationship with the host bacterium; this stable state is known as lysogeny. In a lysogenic cell, phage DNA is normally incorporated into the chromosomal DNA via specific attachment sites on both the phage DNA and the host chromosome. Replication of lambda DNA then occurs only during replication of the host chromosome, and the phage genome is inherited by each daughter cell at cell division. The phage is maintained in this *prophage* state through the action of a repressor protein, coded for by the phage gene cI. This repressor protein turns off the expression of virtually the whole of the lambda genome. If the repressor is inactivated, the expression of phage genes is initiated. This leads to the excision of lambda DNA from the host chromosome and entry into the lytic cycle. The balance between the lytic and lysogenic modes of replication is a delicate and complex one in which a key factor is the concentration of the cI gene product. Some of the many sources of further information about the basic biology of lambda phage are listed in the references to this chapter.

Two methods are described for the large-scale preparation of phage lambda. The first method involves a simple lytic cycle. A culture of a sensitive *E. coli* strain is infected with lambda and incubated, with good aeration, until lysis occurs. The addition of chloroform at the end of this incubation lyses any remaining intact bacteria (including non-infected and lysogenic cells). Phage particles are then concentrated by precipitation with polyethyleneglycol and subsequently purified by cesium chloride gradient ultracentrifugation.

The second preparative procedure described involves the use of a temperature-sensitive lysogen in which the lambda phage carries a mutation (usually cI857) in the repressor gene that renders the repressor protein sensitive to temperatures of 37°C and above. Such a lysogen is therefore grown to a moderate cell density at a reduced temperature (usually 32°C) and is then heat shocked to destroy the function of the phage repressor. This results in the excision of the phage genome and the initiation of the lytic cycle of growth. These cells would then, under normal circumstances, lyse. However, the presence of an additional mutation, usually in phage gene S, prevents lysis of the cells and therefore extends the period of phage production by a particular cell. This results in the accumulation of large numbers of phage particles within the infected cells. These cells can be recovered by centrifugation and subsequently lysed by the addition of chloroform.

This obviates the need for concentrating the phage particles during the subsequent purification procedure. This method is a very simple, convenient procedure for obtaining large quantities of phage for the preparation of DNA for molecular weight markers, for example.

Chapter 30 (preparation of phage DNA) describes an alternative procedure using plate lysates for obtaining comparatively small amounts of phage DNA, either for screening recombinants or for providing the phage inoculum for a large-scale process such a method A.

# Materials

## *For Method A*

1. 500 mL L broth: 7.5 g tryptone; 2.5 g yeast extract; 2.5 g NaCl. Dissolve in about 300 mL of water. Adjust the pH to 7.2 with sodium hydroxide solution and make the volume up to 500 mL with water. Transfer to a 5 L conical flask and sterilize by autoclaving.
2. 1*M* $MgSO_4$; sterilize by autoclaving.
3. Phage inoculum (about $2 \times 10^{11}$ pfu).
4. Chloroform.
5. Deoxyribonuclease, ribonuclease (each at 5 mg/mL in water).
6. Polyethyleneglycol (PEG 6000).
7. Phage buffer:

| | |
|---|---|
| $Na_2HPO_4$ | 7 g |
| $KH_2PO_4$ | 3 g |
| NaCl | 5 g |
| $MgSO_4 \cdot 7H_2O$ | 0.25 g |
| $CaCl_2$ | 0.015 g |
| Gelatin | 0.01 g |
| Water to 1 L | |
| Sterilize by autoclaving | |

8. CsCl solutions for step gradient:

| Density | CsCl, g | Refractive index |
|---|---|---|
| 1.3 | 40 | 1.3625 |
| 1.5 | 67 | 1.3815 |
| 1.7 | 95 | 1.3990 |

Dissolve the cesium chloride in no more than 70 mL of phage buffer, then make up to 100 mL with phage buffer. If a refractometer is available, check the refractive index of the solutions; if not, obtain the density by accurately weighing an aliquot of the solution. Store at 4°C.

### For Method B

1. L broth, 500 mL in a 5 L flask, prepared as described above.
2. SM buffer:

   | | |
   |---|---|
   | NaCl | 0.2*M* |
   | $MgSO_4$ | 0.001*M* |
   | Gelatin | 1% |

   Tris-HCl buffer, pH 7.5 (final concentration 0.05*M*)

   Sterilize by autoclaving. (Note: Phage buffer as described for method A can be used instead).
3. Chloroform.
4. DNase, RNase (each at 5 mg/mL in water).
5. Cesium chloride solutions as for method A.
6. *E. coli* strain: lambda $cI_{ts}$ lysogen, e.g., cI857, Sam7.

### For Assay of Phage

1. 0.01*M* $MgSO_4$.
2. Phage buffer (as described above).
3. BBL agar:

   | | | |
   |---|---|---|
   | Trypticase | 10 g | (Baltimore Biological Laboratories) |
   | NaCl | 5 g | |
   | Agar | 10 g | |

   Add water to 1 L. Heat to dissolve the agar, then dispense in 100 mL amounts and sterilize by autoclaving.
4. BBL top layer agar: as above, but use 6 g of agar/L and dispense it in 3 mL amounts before autoclaving.
5. Sensitive *E. coli* indicator strain.

## Methods

### Method A

1. Dilute 10 mL of a fresh overnight culture of a sensitive strain of *E. coli* (e.g., C600) into 500 mL of warm L

broth in a 5 L flask. Incubate with shaking at 37°C until the optical density at 650 nm is approximately 0.5.

2. Add 5 mL of 1*M* $MgSO_4$ and approximately 2 × $10^{11}$ pfu of the required phage. Continue incubation at 37°C with shaking until the culture lyses. This should take about 2–3 h. Lysis may be indicated by a rapid drop in optical density, but do not be too surprised if there is no visible clearing of the culture. A fully lysed culture will contain substantial amounts of bacterial debris that may appear quite turbid; the nature of the turbidity will change, however, and in a lysed culture there may be stringy patterns of DNA and protein or larger clumps of cell debris.
3. Add 1 mL of chloroform and shake for a further 15 min.
4. Add 20 g of NaCl and mix to dissolve.
5. Add DNase and RNase (final concentration 1 µg/mL) and leave at room temperature for 1 h.
6. Remove the bacterial debris by centrifugation at 15,000*g* (10,000 rpm in a 6 × 250 mL rotor) for 10 min at 4°C.
7. Take the supernatant and add 50 g of solid PEG 6000. Mix to dissolve, using a magnetic stirrer, and allow the phage particles to precipitate at 4°C for several hours, or (preferably) overnight.
8. Centrifuge the suspension at 15,000*g* (10,000 rpm in a 6 × 250 rotor) for 10 min. Carefully remove and discard the whole of the supernatant, allowing the pellet to drain by inverting or tilting the centrifuge pots. Gently resuspend the pellet in 10 mL of phage buffer by shaking at 4°C for 30–60 min, or by gentle pipeting with a wide-bore pipet. It is important to do this gently to avoid shearing off the phage tails, which will result in non-infectious particles.
9. Clarify the suspension by centrifugation at 15,000*g* for 5 min.
10. Form a cesium chloride step-gradient in a Beckman SW41 centrifuge tube (or equivalent) by adding first 1.5 mL of the 1.7 g/mL solution, then carefully layering, in turn, 2 mL of the 1.5 g/mL solution and 2 mL of the 1.3 g/mL solution. The clarified phage suspension is then layered onto the step gradient.

11. Centrifuge at 180,000*g* (38,000 rpm in a Beckman SW41 rotor) for 2–3 h at 15°C.
12. A turbid band of phage particles should be clearly visible at the interface between the 1.3 and 1.5 g/mL layers. This band can be collected by piercing the side of the tube with a 21G syringe needle, just below the band. Sticking a piece of sellotape to the tube before piercing with the needle can help to prevent leaks; make sure the position of the sellotape does not obscure the phage band.
13. Transfer the phage suspension to a smaller ultracentrifuge tube (e.g., for Beckman SW50.1 rotor). Fill the tube with CsCl solution (density 1.5 g/mL in phage buffer) and centrifuge to equilibrium (36–48 h) at 120,000*g* (35,000 rpm in a Beckman SW50.1 rotor) at 10°C.
14. Collect the phage band as before. Transfer to a tightly capped tube or ampule and store, in cesium chloride, at 4°C. In this form, the phage can be stored indefinitely.

## Method B

1. Inoculate 5 mL of a fresh overnight culture of the lambda lysogen into 500 mL of warm L broth in a 5 L flask and shake at 32°C.
2. Measure the optical density at 650 nm at 30 min intervals. When the optical density reaches 0.45, transfer the culture to a 42°C water bath. Induce the lysogen by shaking at 42°C for 30 min; note that a rapid increase in temperature is needed for good phage induction. Finally, return the culture to 37°C and shake for a further 1–3 h.
3. If the phage carries a wild-type S gene, lysis of the induced bacteria will occur spontaneously. In that case, add 10 mL of chloroform to complete lysis and proceed as for Method A.
4. If the phage is defective in gene S, lysis will not occur spontaneously and the phage particles will remain within the bacterial cells. These cells can be recovered by centrifugation at 9000*g* (8000 rpm in a 6 × 250 mL rotor) for 10 min and can then be stored at −20°C before further processing.

5. Resuspend the cell pellet in 20 mL of SM buffer and transfer to a 30 mL glass (corex) centrifuge tube.
6. Add 2 mL of chloroform and gently stir the bacterial suspension with a glass rod or pipet. Cell lysis will be indicated by an increase in the viscosity of the suspension. Continue stirring for about 15 min.
7. Add 200 μL of DNase (5 mg/mL) and 200 μL of pancreatic RNase (5 mg/mL) and incubate at room temperature for 1–2 h. The viscosity of the suspension should be greatly reduced.
8. Remove the bacterial debris by centrifugation at 7000*g* (8000 rpm in an 8 × 50 mL rotor) for 10 min. Recover the supernatant and store it on ice. The yield of phage can be increased by washing the pellet with 2 mL of SM buffer, recentrifuging, and combining the two supernatants.
9. Phage particles present in the supernatant, which should be slightly turbid, can then be purified by a cesium chloride step gradient as described for method A.

## *Method for Assay of Phage*

1. Grow an overnight culture of the sensitive indicator strain in L broth.
2. Pellet the cells in a bench centrifuge and resuspend them in half the culture volume of magnesium sulfate solution. These cells can be stored for several weeks at 4°C. (**Note**: this step is optional if the cells are to be used fresh; the overnight cells can then be used for plating phage without any further treatment and without any great reduction in plating efficiency).
3. Make serial dilutions of the phage in phage buffer. Mix 100 μL of diluted phage with 100 μL of plating cells and incubate at 37°C for 15 minutes to pre-absorb the phage to the plating cells.
4. Melt the required number of bottles of top-layer agar in a boiling water bath, and cool them to 45°C in a water bath.
5. Dry the outside of a bottle of top-layer agar, add one of the phage-cell mixtures, and pour onto a dry BBL plate.
6. Allow the agar to set and incubate the plates overnight at 37°C.

7. Score the number of plaques and calculate the phage titer in plaque-forming units (pfu)/mL. Temperate phages such as wild-type lambda will give turbid plaques owing to the growth of lysogenized bacteria within the plaque. Virulent phages (and cI mutants of lambda) will give clear plaques.

## Notes

1. In principle, the above methods can be applied to other bacteriophages, with Method A being applicable to virulent as well as temperate phages. However, the properties of phages vary considerably, so it is *essential* to check the literature before proceeding. For example, some phages are rapidly inactivated by chloroform or by cesium chloride. The assay conditions (medium, temperature of incubation, etc.) may also vary.
   Note that some temperate phages can be induced by alternative treatments, e.g., mitomycin C or ultraviolet irradiation.
2. When carrying out assays, it is essential to make sure that the indicator strain is sensitive to the phage being assayed. This means it must not only be capable of being infected, but that infection must result in the formation of plaques. In particular, if the phage carries an amber mutation in gene S, the indicator strain must be an amber suppressor. The indicator should also be deficient in the *E. coli* restriction system. If an $r^-$ mutant is not available, then it is best to use as an indicator the same *E. coli* strain as that used to grow the phage in the first place. This is very important. Different strains of *E. coli* carry different restriction and modification systems; phages with unmodified DNA will show an extremely low efficiency of plating.
3. Contamination of the plates is often caused by the water from the water bath in which the top layer agar bottles are cooled. It is essential that the bottles be dried before pouring.
4. If large numbers of assays are to be done, the overlays can be prepared in larger volumes and dispensed using an automatic dispenser.

5. Large numbers of dilutions can be conveniently carried out using sterile microtiter plates.
6. It is essential that the overlays should be spread evenly over the plates before they start to set. This can be facilitated by warming the plates in a 37°C incubator before they are used.
7. Pre-absorption of the phage to the plating cells improves the results, but is not absolutely essential. The procedure can therefore be simplified by adding the phage dilution and the plating cells directly to the overlay.
8. A fresh lambda plaque contains between $10^5$ and $10^7$ plaque-forming units.
9. The addition of maltose to the medium in which the indicator cells are grown can give improved results owing to the consequent induction of the lambda receptor protein. This also applies to Method A for the preparation of the phage. Prepare and sterilize a 20% solution of maltose in distilled water, and add 1 mL of this solution to each 100 mL of growth medium.

# References

1. Brammar, W. J. (1982) Vectors based on bacteriophage lambda. In *Genetic Engineering,* volume 3 (Williamson, R., ed.) pp. 53–81. Academic Press, London.
2. Hendrix, R. W., Roberts, J. W., Stahl, F. W., and Weisberg, R. A. (1982) *Lambda II.* Cold Spring Harbor Laboratory, NY, USA.
3. Weisberg, R. A., Gottesman, S., and Gottesman, M. E. (1977). Bacteriophage lambda: the lysogenic pathway. In *Comprehensive Virology* vol 8 (Fraenkel-Conrat, H., and Wagner, P. R., eds.) pp. 197–258. Plenum, New York.
4. Williams, B. G., and Blattner, F. R. (1980) Bacteriophage lambda vectors for DNA cloning. In: *Genetic Engineering,* (Setlow, J. K., and Hollaender, A., eds.) pp. 201–229. Plenum, New York.
5. Yamamoto, K. R., Alberts, B. M., Benzinger, R., Lawhorne, L., and Treiber, G. (1970) Rapid bacteriophage sedimentation in the presence of polyethylene glycol and its application to large-scale virus purification. *Virology* **40,** 734–744.

# Chapter 30

# Preparation of Phage Lambda DNA

*J. W. Dale and P. J. Greenaway*

*Department of Microbiology, University of Surrey, Guildford, Surrey and Molecular Genetics Laboratory, PHLS Centre for Applied Microbiology and Research, Porton Down, Salisbury, Wilts., United Kingdom*

## Introduction

The isolation of phage DNA from a purified phage preparation simply involves the removal of the proteins of the phage particle. This is done most easily by phenol extraction (Method A). The large amounts of high quality DNA needed for use as vectors or for markers for gel electrophoresis can be obtained in this way. However, sometimes (for example, when screening recombinant phages) only small amounts of phage DNA are required. The purity of these samples needs only to be sufficient for the restriction enzyme digests to provide distinct patterns on an agarose gel. Method B provides one way in which this can be done (*1*). A plate lysate is prepared and the released phage is subsequently lysed by the addition of SDS. The phage proteins and SDS are precipitated by the addition of potassium acetate to leave the phage DNA in solution. Finally, the DNA is recovered by ethanol precipitation.

## Materials

### *For Method A—Using Large-Scale Lysates*

1. Purified phage preparation, obtained as described in Chapter 29.
2. TE buffer: 10 m*M* Tris-HCl, 1 m*M* EDTA, pH 8.0
3. Phenol: freshly distilled, neutralized, and equilibrated with TE buffer.

### *For Method B—Using Plate Lysates*

1. Phage buffer, prepared as described in Chapter 29.
2. L-agarose plates: L broth solidified with 10 g/L of agarose.
3. BBL agarose top layers: prepared as for BBL top layers (see Chapter 29) but using 6 g/L of agarose instead of agar. (There is often contaminating material present in agar that inhibits restriction enzymes).
4. Overlay buffer: 10 m*M* Tris-HCl, 10 m*M* EDTA, pH7.5.
5. 0.5*M* EDTA.
6. 2*M* Tris base.
7. 10% SDS (w/v).
8. Diethylpyrocarbonate (DEP).
9. 5*M* potassium acetate.
10. Absolute ethanol.
11. TE buffer: 10 m*M* Tris-HCl, 1 m*M* EDTA, pH 8.0

# Methods

## Method for Large-Scale Lysates

1. If the phage preparation has been stored in cesium chloride, it must be dialyzed against four changes (each of 500 mL) of TE buffer at 4°C.
2. After dialysis, add an equal volume of phenol and shake gently for 2–3 min.
3. Centrifuge briefly and remove the lower (phenol) phase. Do this at least three times. Chill the tube in ice after the final phenol extraction (and removal of the phenol layer) centrifuge and take the aqueous layer.
4. Transfer the aqueous phase to dialysis tubing and dialyze against four changes (500 mL each) of TE buffer at 4°C.
5. Transfer the DNA to a sterile tube and store at 4°C. The ratios of absorbance at 260:235 and 260:280 should be greater than 1.7.

## Method for Plate Lysates

1. Pick a single plaque with a Pasteur pipet and transfer to a tube containing 1 mL of phage buffer. Incubate at 4°C for several hours to allow the phage to diffuse out of the soft agar plug. This will provide a phage suspension containing about $10^5$ pfu/mL.
2. Mix 0.1 mL of this suspension with 0.1 mL of a fresh overnight culture of a sensitive indicator strain. Incubate for 20 min at 37°C to allow absorption of the phage to the bacteria. The remainder of the phage suspension can be kept at 4°C as a temporary stock for that clone.
3. Add 3 mL of BBL top layer agarose (melted and cooled to 45°C) and pour onto an L-agarose plate. Incubate at 37°C until confluent lysis is just observed (6–8 h).
4. Overlay the plate with 5 mL of the overlay buffer. If this procedure is intended to produce viable phage particles rather than DNA, use phage buffer instead.

Incubate at 4°C for several hours, or overnight if possible.

5. Recover 4 mL of the overlay buffer, add 0.4 mL of EDTA, 0.2 mL of Tris, and 0.2 mL of SDS. Mix and leave on ice for 5 min.
6. Add 10 μL of DEP and heat at 65°C for 30 min in an open tube in a fume cupboard. The purpose of this step is to prevent nuclease action.
7. Place the solution, which should now contain completely lysed phage particles, in an ice bath and add 1 mL of 5*M* potassium acetate. Leave on ice for 1 h.
8. Remove the white precipitate by centrifugation at 15,000 rpm for 10 min.
9. Recover the supernatant and add 11 mL of absolute ethanol. Mix the solution thoroughly and leave at −20°C overnight, or at −70°C for 30 min.
10. Recover the precipitated nucleic acids by centrifugation at 30,000*g* for 30 min. Remove all the ethanol carefully and dry the DNA in a vacuum desiccator. Resuspend the DNA in 100 μL of TE buffer.
11. 10 μL of this DNA solution is usually sufficient to give distinct bands on agarose gels after restriction enzyme digestion. However, the yields are variable. Since this procedure gives nicked DNA, heat-shocking restriction enzyme digests may cause the DNA to disintegrate and should therefore be avoided. The storage properties of DNA prepared in this way are poor, but may be improved by phenol extraction and a second ethanol precipitation.

# Notes

## *For Large-Scale Lysates*

1. Approximately $2 \times 10^{10}$ phage particles are required for 1 μg of DNA.
2. Some intact phage can be carried through this procedure.

## References

1. Cameron, J. R., Philippsen, P., and Davis, R. W. (1977) Analysis of chromosomal integration and deletions of yeast plasmids. *Nucleic Acid Res.* **4,** 1429–1448.

# Chapter 31

# The Use of Restriction Endonucleases

***Elliot B. Gingold***

*School of Biological and Environmental Sciences, The Hatfield Polytechnic, Hatfield, Hertfordshire, England*

## Introduction

The discovery of the mode of action of the class of bacterial enzymes known as restriction endonucleases provided the major breakthrough in opening up the field of genetic engineering. In vivo, these enzymes are involved in recognizing and cutting up foreign DNA entering the cell; their most likely role is thus protecting the bacteria against phage infection. The property that is relevant to us is that these enzymes recognize specific DNA sequences. The enzymes used in DNA manipulations are in fact known as Class II restriction endonucleases; these enzymes cut the DNA within the recognition sequence at a defined point. Treatment of a DNA sample with such enzymes will thus result in each molecule being cut at the same positions and thereby lead to the formation of reproducible fragments.

In Table 1 a number of commonly used restriction enzymes are listed along with their recognition sequences. It can be observed that a wide range of recognition sequences are available. In general, the simpler the sequence is, the more likely it is to be found in a given DNA molecule, and hence the more cuts that the enzyme will make.

Table 1
Properties of Some Common Restriction Endonucleases

| Enzyme | Recognition sequence[a] | Buffer[b] |
|---|---|---|
| AccI | GT ↓ $\begin{pmatrix} A & G \\ C & T \end{pmatrix}$ AC | Med |
| AluI | AG ↓ CT | Med |
| AvaI | C ↓ Py CGPu G | Med |
| BamHI | G ↓ GATCC | Med |
| Bcl1 | T ↓ GATCA | Med (60°C) |
| Bgl11 | A ↓ GATCT | Low |
| EcoRI | G ↓ AATTC | High |
| EcoRII | ↓ CC$\begin{pmatrix} A \\ T \end{pmatrix}$ GG | High |
| HaeIII | GG ↓ CC | Med |
| HhaI | GCG ↓ C | Med |
| HincII | GT ↓ Py PuAC | Med |
| HindIII | A ↓ AGCTT | Med |
| HinfI | G ↓ ANTC | Med |
| HpaI | GTT ↓ AAC | Low |
| KpnI | GGTAC ↓ C | Low |
| MboI | ↓ GATC | High |
| MspI | C ↓ CGG | Low |
| PstI | CTCGA ↓ G | Med |
| PvuII | CAG ↓ CTG | Med |
| SacI | G AGCT ↓ C | Low |
| SalI | G ↓ TCGAC | High |
| Sau3A | ↓ GATC | Med |
| TaqI | T ↓ CGA | Low |
| XbaI | T ↓ CTAGA | High |
| XhoI | C ↓ TCGAG | High |
| XmaI | C ↓ CCGGG | Low |

[a]Base sequences are shown 5′ to 3′. Where two bases are shown at a position, these are alternatives. Pu represents either purine, Py either pyrimidine, N any base.
[b]Refers to high, medium, and low salt buffers (*see* Materials).

It should also be noted that the position of the cuts may be in the center of the sequence (leading to blunt ends) or to one side (leading to single-stranded tails or cohesive ends).

There are two purposes for which restriction endonucleases are commonly used. They may be used to provide "landmarks" on a physical map of a DNA molecule. In this process, known as restriction mapping, the aim is to determine the positions on the molecule of the recognition sequences of a number of enzymes. Secondly, cutting a molecule with such enzymes the basis of obtaining the DNA fragments to be used in a genetic manipulation. In both types of procedure the basic method is the same although the purity of the enzyme preparation is more critical if preparing samples for subsequent ligation than if merely analyzing fragment sizes by gel electrophoresis.

With the wide range of commercially available enzymes, it is becoming less common for laboratories to prepare their own supplies of enzymes. One initially daunting factor is, however, that each source of enzyme provides a different recipe for the optimum incubation buffer. In fact, most enzymes will work well in one or other of three standard buffers, the major difference being in the overall ionic strength. Table 1 includes the buffer preferences for the enzymes listed, it is a simple matter to deduce from the recipes supplied which buffer system is appropriate when working with other enzymes. Do note, however, that some enzymes will have special requirements, such as a pH optimum other than 7.5 or a temperature optimum other than 37°C. Nonetheless, for most enzymes it is convenient to use one of the standard buffers rather than producing a separate manufacturer's recommended buffer for each.

## Materials

1. Buffers (made up as listed here being at ×10 final concentration. Store stock solutions at −20°).

| | |
|---|---|
| (a) Low salt | 100 m*M* Tris (pH 7.5), 100 m*M* $MgCl_2$, 10 m*M* dithiothreitol |
| (b) Medium salt | 0.5*M* NaCl, 100 m*M* Tris (pH 7.5), 100 m*M* $MgCl_2$, 10 m*M* dithiothreitol |
| (c) High salt | 1*M* NaCl, 0.5*M* Tris (pH 7.5), 100 m*M* $MgCl_2$, 10 m*M* dithiothreitol |

2. Sterile distilled water, carefully prepared to avoid traces of detergent in bottles.
3. DNA Samples. For fragment size analysis it is desirable to include tubes with a known standard such as λDNA.
4. Sterile 0.5 mL Eppendorf tubes, micropipet tips.
5. Stopping Mix:

   (a) For gel electrophoresis: 20% sucrose, 10% Ficoll, 100 m*M* EDTA, 1% bromophenol blue
   (b) For purification and ligation: 100 m*M* EDTA

# Method

1. Prepare a clear protocol showing what is to be added to each tube. For each tube, use a total reaction volume of 20 μL.
2. An individual tube should contain 0.2–1 μg of DNA (*see* Note 1), 2 μL of the concentrated buffer, one or more units of the enzyme (*see* Note 2), and sterile distilled water to make up the volume. The amount of water added will depend on the concentration of the initial DNA solution and of the enzyme preparation. ALL volumes should be calculated before commencing.
3. To each tube add the appropriate quantity of DNA solution, sterile distilled water, and buffer. When all tubes are ready, obtain the enzyme preparation from the freezer, withdraw the appropriate amounts, add to the tubes, and return enzyme to freezer. Since most preparations are stored in 50% glycerol, no thawing is necessary. Do not leave the preparation out of the freezer longer than is necessary and use a fresh pipet tip for each withdrawal.

4. Mix well by tapping the tube and then spin down the contents for 1 s in a microcentrifuge. Since incomplete mixing can often reduce digestion, it is worth repeating this process.
5. Incubate the tubes for 1 h (or whatever interval is required) in a 37°C water bath.
6. To each tube add 2 μL of stopping mix. Repeat Step 4. The sample is now ready for analyses by gel electrophoresis (*see* Chapter 7) or for deproteinization followed by ligation (*see* Chapter 32).

# Notes

1. The quantity of DNA to be digested will depend on the purpose for which it is required. If the digested sample is to be analyzed by gel electrophoresis the amount required for loading will vary with the nature of the DNA. Thus with a plasmid that is cut once by an enzyme, 0.2 μg of DNA may produce an overloaded band, whereas 0.2 μg of phage λ cut with EcoRI or Hind III will produce a clear banding pattern. With a DNA of greater complexity (and hence larger numbers of bands), a larger quantity of DNA may be needed for clear band observation.

   It is generally desirable, however, to keep the volume of DNA solution added relatively small. This is done so that the other components of the DNA solution do not influence the reaction or the running characteristics on the gel.
2. Enzyme activity is generally described in terms of units, by which 1 unit of an enzyme will completely digest 1 μg of λ in 1 h. It would seem from this definition that unless the DNA in question is especially rich in recognition sequences for the enzyme, it should be sufficient to add 1 unit of enzyme. This can, however, be misleading. Many DNA preparations will not be as susceptible to digestion as λ, possibly because of inhibition by trace contaminants. In addition some sites can be particularly resistant to digestion. In general it may be worthwhile using a range of enzyme concentrations to

determine the required quantity. It is not advisable to use great excesses of enzymes since, apart from cost, exonuclease impurities in the enzyme preparation may reduce the ability of the DNA to religate.

3. Enzymes are often supplied in very concentrated form and to avoid measuring very small volumes it can be useful to dilute them in the reaction buffer (×1). If this is to be done, it is convenient to use a dilution factor that gives the required activity in 2 μL.
4. Increasing the time of reaction is an alternative to increasing the enzyme concentration. Some enzymes, however, will lose activity quite rapidly and, hence, this will not always be successful.
5. It is worth remembering that bacteria protect their own DNA by methylation of bases within the recognition sequence. It is thus apparent that methylation patterns on the DNA sample could protect some sites from cutting. Although this can be a problem, it also provides a basis for studies of levels of base modification in eukaryotic genomes.
6. For purposes of restriction mapping, it is generally necessary to subject samples to digestion with two enzymes. If the two enzymes have similar requirements, this can be done simultaneously. In fact, sometimes the wide range of acceptable conditions allows unexpected success. For example, EcoRI and Hind III work well together in high salt buffer. Do note, however, that the wrong conditions can produce misleading results. Thus, in low salt buffers, EcoRI produces many additional cuts at sites that have only the inner four bases (AATT).

   If the same buffer cannot be used by each enzyme, the reactions are performed sequentially. A number of approaches are possible:

   (a) The reaction requiring the lower salt buffer can be performed first, extra salt added and the second enzyme then introduced.
   (b) The first reaction can be performed in a small volume, the sample then being diluted into the buffer required by the second enzyme.

(c) After the first reaction the DNA can be purified by extracting with an equal volume of phenol/chloroform followed by precipitation of the DNA in 5*M* ammonium acetate and 2 vol ethanol. The tube is left at −70°C for 30 min, then centrifuged in an Eppendorf microfuge for 5 min. After drying the pellet is made up in the second buffer.

With each method, analysis of a small sample of the reaction mixture by gel electrophoresis on a mini-gel system provides a useful indication of the completeness of each stage.

7. It is also often required to perform partial digestions, cutting at only a fraction of the possible sites. A range of enzyme concentrations can be used and the results monitored by analyzing small samples on mini gels. Variation in time of reaction can also be used to vary the degree of digestion.

# Chapter 32

## Ligation of DNA with $T_4$ DNA Ligase

***by Wim Gaastra and Kirsten Hansen***

*Department of Microbiology, The Technical University of Denmark, Lyngby, Denmark*

## Introduction

Since they are involved in such important processes as DNA replication, DNA repair, and DNA recombination, DNA ligases can be found in all living cells. Two prokaryotic DNA ligases have become indispensible tools in the fields of in vitro DNA recombination and DNA synthesis. DNA ligase from *E.coli* is a polypeptide with a molecular weight of 74,000 and is NAD-dependent. $T_4$ DNA ligase is the product of gene 30 of the $T_4$ phage, has a molecular weight of 68,000, and is ATP-dependent. Both enzymes catalyse the synthesis of a phosphodiester bond between the 3′-hydroxyl group and the 5′-phosphoryl group at a nick in double-stranded DNA. Of the two enzymes, only $T_4$ DNA ligase is able to join both DNA fragments with protruding ("sticky ends") as well as DNA fragments with "blunt ends" (*1*). The reaction takes place in three steps. The first step is the transfer of an adenylyl group of ATP (for $T_4$ ligase) or NAD (for *E. coli* ligase) to the free enzyme. It is a side chain $NH_2$ group of a lysine residue that becomes adenylated. During this step, pyrophosphate ($T_4$ enzyme) or nicotinamide monophosphate (*E. coli* enzyme) are released. In the second step, the adenylyl group is then transferred to the 5′-phosphoryl end of the DNA and in the third step a phosphodiester bond is formed between a 3′-hydroxyl group and the 5′-adenylated phosphoryl group with the release of AMP. [*See* also ref. (*5*) for more detail.]

## Materials

1. 100 m*M* Tris-HC1, pH 7.8; 20 m*M* $MgCl_2$.
2. 4 m*M* ATP (sterilize by filtration).
3. 80 m*M* dithiothreitol (DTT) (sterilize by filtration).
4. $T_4$ DNA ligase (Boehringer 0.9 U/μL).
5. 3*M* NaAc.
6. 96% Ethanol.

## Method

1. Prepare the Tris-buffer, ATP, and DTT solutions described under materials, and divide them as follows in separate Eppendorf tubes: 100 μL of Tris-buffer per tube, 50 μL of ATP solution per tube, 50 μL of DTT solution per tube. These tubes are kept in the freezer at −20°C until further use.
2. Mix the two solutions of the DNA fragments to be ligated (*see* Note 3) and precipitate with Ethanol by adding 0.1 vol of 3*M* NaAc and 3 vol of 96% ethanol. Mix and precipitate at −70°C for 10 min or at −20°C for 30 min.
3. Centrifuge in an Eppendorf centrifuge for 10 min, then discard the supernatant and air dry the pellet.

4. Take one tube with Tris-buffer, one with ATP, and one with DTT, mix them, and use 20 μL of the mixture for each ligation reaction.
5. Dissolve the air-dried DNA in the 20 μL of ligation mixture and add 0.5 μL of $T_4$ DNA ligase. Mix by gently tapping the tube. (Do not mix on a whirl mixer, to prevent denaturation of the $T_4$ ligase.) Then incubate overnight at 15°C (*see* Notes 6 and 7).
6. If necessary, following the incubation, check the ligation reaction on an agarose gel. Run nonligated DNA fragments in parallel tracks as controls.

   If the ligation has proceeded sufficiently, proceed with the next step, which in most cases will be transformation of competent cells (*see* Chapters 34 and 35).

# Notes

The protocol given above is a general protocol for both blunt ends and cohesive end ligation. The effectiveness of the reaction can, however, be increased by taking into account some of the following notes.

1. The amount of DNA ligase needed to ligate some DNA fragments depends on the nature of the DNA to be ligated, especially whether it is blunt ends or protruding ("sticky") ends that should be ligated.

   Generally 10–100 times more enzyme has to be used for the ligation of blunt ends than for sticky end ligation.
2. The ligation of blunt ends is not facilitated by alignment caused by base pairing and is remarkably stimulated by $T_4$ RNA ligase (2), although the enzyme itself does not join blunt ends. Especially at low $T_4$ DNA ligase concentrations, as much as a 20-fold increase in ligation can be obtained in the presence of $T_4$ RNA ligase. This stimulation is specific for blunt ends; no influence was found with sticky ends.
3. When DNA fragments with cohesive ends are to be joined, DNA concentrations between 2 and 10 μg/mL are used. When DNA fragments with blunt ends are to be ligated, the concentration is between 100 and 200

μg/mL. (Concentrations are calculated for fragments with an average length between 200 and 5000 base pairs long.) Reactions in which an intermolecular product is desired should be carried out at high DNA concentrations. Circularization of DNA is favored by low concentrations and short-length DNA molecules.

4. Several assays have been used to define the activity of ligases. The most practical one, for in vitro recombination, is the determination of the amount of enzyme needed to religate 50% of the HindIII fragments of phage lambda under standard conditions. But it is recommended to carefully read the manufacturers specification for the amount to use in every case.
5. After ligation the mixtures can be stored for several days at 0°C and for 2 months at −8°C. They should however not be frozen at −20°C, since material thus treated gives a low efficiency of transformation.
6. The hydrogen bonds between the cohesive ends of the various DNA fragments to be ligated are more stable at low temperatures than at higher (4). Therefore cohesive ends are usually ligated at temperatures between 10 and 16°C, although the optimal temperature of $T_4$ DNA ligase is probably 37°C. However, when high DNA concentrations of short DNA fragments with strong cohesive ends are used, higher temperatures might be used also.
7. Ligation of blunt ends is generally performed at slightly higher temperatures (15–20°C) and for longer periods (6–18 h) than ligation of cohesive ends since the chance that two DNA ends meet and remain aligned long enough to be joined is very small.
8. Ligation by $T_4$ DNA ligase of blunt ends is completely inhibited at ATP concentration of 5 m*M*. Cohesive end ligation is only inhibited by 10% (3). The optimal ATP concentration for both blunt ends and cohesive end ligation is between 0.5 and 1 m*M*.
9. When a mixture of DNA fragments with blunt and cohesive ends is used it is possible to ligate selectively only the cohesive ends by using *E.coli* DNA ligase, since this enzyme does not join blunt ends or by making use of the observations in Note 8.

10. One of the nuisances encountered when cloning with plasmids is that most transformants contain only the recircularized cloning vehicle in stead of the cloning vehicle with an insert. Since 5′-hydroxyl groups do not act as substrate for $T_4$ DNA ligase, this fact can be used to prevent recircularization of the cloning vector, by removing the terminal 5′-phosphate groups from the cloning vehicle [with alkaline phosphatase] prior to ligation, (*see* Chapter 33). This results in a very low frequency of transformation of cloning vehicles because of the poor ability of linear DNA to undergo in vivo circularization. However, the 5′-phosphate ends of the DNA fragments to be cloned can be ligated to the 3′-hydroxyl ends of the cloning vehicle. This results in circular molecules of vector and passenger DNA, with single stranded nicks at the ends of the vector fragment. These molecules transform as efficiently as completely covalently closed circular DNA. The nicks are repaired in vivo after transformation. Note that a higher concentration of DNA ligase is needed for complete ligation of the DNA, than if the DNA was not treated with alkaline phosphatase.
11. Generally, when the cloning vehicle is not treated with alkaline phosphatase, an excess (2–5-fold) of DNA fragments to be cloned over vector molecules is used. This assures that the cloning vehicle preferentially form intermolecular bonds with the DNA fragments to be cloned, rather than recircularized cloning vehicle.

# References


1. Sgaramella, V., van de Sande, J. H., and Khorana, H. G. (1975) Studies on polynucleotides. A novel joining reaction catalyzed by the $T_4$ polynucleotide ligase. *Proc. Natl. Acad. Sci. USA* **67,** 1468–1475.
2. Sugino, A., Goodman, H. M., Heyneker, H. L., Shine, J., Boyer, H. W., and Cozzarelli, N. R. (1977) Interaction of bacteriophage $T_4$ RNA and DNA ligases in joining of duplex DNA at base paired ends. *J. Biol. Chem.* **252,** 3987–3994.

3. Ferretti, L., and Sgaramelle, V. (1981) Specific and reversible inhibition of the blunt end joining activity of the $T_4$ DNA ligase. *Nucleic Acid Res.* **9,** 3695–3705.
4. Dugaiczyk, A., Boyer, H. W., and Goodman, H. M. (1975) Ligation of EcoRI endonuclease generated DNA fragments into linear and circular structures. *J. Mol. Biol.* **96,** 171–184.
5. Mooi, F. R., and Gaastra, W. (1983) The use of restriction endonucleases and $T_4$ DNA ligase, in *Techniques in Molecular Biology,* Walker, J. M., and Gaastra, W., eds. Croom Helm, London, pp. 197–219.

# Chapter 33

# The Use of Alkaline Phosphatase to Prevent Vector Regeneration

**_J. W. Dale and P. J. Greenaway_**

*Department of Microbiology, University of Surrey, Guildford, Surrey and Molecular Genetics Laboratory, PHLS Centre for Applied Microbiology and Research, Porton Down, Salisbury, Wilts., United Kingdom*

## Introduction

Linearized plasmid vector molecules produced by restriction enzyme digestion can be reformed into circles by the action of DNA ligase. This is the most favorable reaction even when foreign DNA fragments are present. Thus, unless a direct selection technique is available, the reformed parental molecule will also give transformants and hence reduce the overall cloning efficiency. Usually most of the recircularized parental plasmids can readily be distinguished from recombinants by the absence of insertional inactivation; nevertheless, this may involve screening large numbers of transformants for each recombinant.

The seriousness of this problem can be reduced by a careful choice of ligation conditions (*see* Chapter 32 on DNA ligation). For example, keeping the volume of the ligation mixture as small as practicable (and hence keeping the DNA concentration high) will favor intermolecular rather than intramolecular ligation. Using an excess of the insert DNA will also favor the formation of recombinants, but may also lead to random ligation of foreign DNA and multiple insertion.

A more reliable strategy is to treat the cleaved vector DNA with alkaline phosphatase (*1*) that removes the 5′-phosphate group from the DNA. Since the presence of a terminal 5′-phosphate is necessary for the action of DNA ligase, a dephosphorylated DNA strand will not be able to participate in a ligation reaction. Since this will apply to both ends of the linearized vector, the ends cannot be joined together again, i.e., parental vector regeneration is impossible. This results in a very low frequency of transformation by the vector (since linear DNA transforms very poorly). However, the vector 3′-hydroxyl ends *are* available for joining (by DNA ligase) to the 5′-phosphate termini of any DNA fragments to be cloned. This will result in circular recombinant molecules with a single-stranded nick at each end of the vector portion. These molecules usually transform as efficiently as the fully ligated product, and the nicks are repaired in vivo after transformation. Thus the phosphatase treatment of the vector molecules increases the proportion of hybrid DNA molecules in the population of transformed cells, and also prevents the formation of oligomeric forms of the vector.

## Materials

1. Enzyme: Alkaline phosphatase from calf intestine. This should be free of DNase, phosphodiesterase, and nicking activity, but may contain small amounts of RNase. If the presence of RNase is likely to cause problems, it can be removed by Sephadex gel filtration (*2*). If the enzyme is supplied as an ammonium sulfate sus-

pension, mix gently, remove 10 U to a microfuge tube, and centrifuge. Discard the supernatant and resuspend the pellet in 200 μL of phosphatase dilution buffer. This forms the stock solution, which should be stored at −20°C; it is then stable for at least 6 months, probably considerably longer.

2. Phophatase dilution buffer: 100 m*M* glycine-HCl (pH 9.5), 1 m*M* $MgCl_2$, 0.1 m*M* $ZnCl_2$, all in 50% glycerol.
3. Reaction buffer: 100 m*M* glycine-HCl (pH 9.5), 1 m*M* $MgCl_2$, 0.1 m*M* $ZnCl_2$.

   This buffer is used for optimum activity of the enzyme, but for convenience, the enzyme has sufficient activity in most restriction enzyme buffers to be added directly to the restriction digest. For example, the activity on the PNP substrate in Eco R1 buffer is approximately 60% of the maximum rate.
4. Assay substrate: *p*-nitrophenyl phosphate (PNP). Dissolve 5 mg in 1.2 mL of water and store at −20°C.
5. Other reagents: phenol (freshly neutralized), ethanol, dry ice.

# Method

1. Digest the DNA with restriction enzyme(s) in the usual manner, using approximately 1 μg DNA/20 μL of reaction mix. Terminate the reaction by heat treatment (65°C, 10 min). Note that some restriction enzymes cannot be killed by heat treatment.
2. Add 5 μL of the stock phosphatase and incubate at 37°C for 60 min.
3. At the end of this time, the activity of the enzyme can be checked by adding stock PNP to the level of 5% by volume; the reaction mixture should then quickly turn yellow.
4. Since phosphatase cannot be inactivated by heat, phenol extraction is necessary. Add an equal volume of phenol, mix by repeated inversion, and separate the layers by centrifugation. Using a drawn-out Pasteur pipet, carefully remove the lower (phenol) layer. Repeat the phenol treatment and removal of the phenol layer, then chill the tube, centrifuge, and take the upper

(aqueous) layer to a fresh tube. The phenol extraction will also remove the yellow PNP product.

5. Traces of phenol can be removed, if necessary, by extraction with water-saturated ether. However, the ethanol precipitation step is usually sufficient.
6. Precipitate the DNA (and remove the phosphate liberated by the phosphatase treatment) by ethanol precipitation. Add 0.1 vol 3*M* ammonium acetate plus 2.5 vol chilled ethanol, place the tube in a dry ice–ethanol bath for 10–15 min, and centrifuge in the cold. Repeat the freezing and centrifugation steps before carefully removing the supernatant with a drawn-out Pasteur pipet.
7. Wash the precipitated DNA three times with cold (−20°C) ethanol, and then dry in a vacuum desiccator. Redissolve DNA in a minimum volume of the buffer of choice prior to ligation.

## Notes

1. The most valid test for the effectiveness of the above procedure is the abolition of self-ligation of the DNA. The following control transformations can therefore be performed:

   (a) Restriction enzyme-digested vector DNA
   (b) Digested vector DNA with phosphatase treatment
   (c) Digested vector DNA, phosphatase treated, and ligated
   (d) Digested vector DNA, ligated without previous phosphatase treatment
   (e) Intact vector

   Transformation (c) should give results at a level comparable with (often lower than) that of linear DNA (a) and (b), and much lower than that of the recircularized vector (d). Transformation with the religated vector should be very much higher than with the unligated

DNA, but is not expected to approach the level seen with the intact, undigested vector DNA.

2. Alkaline phosphatase will effectively dephosphorylate protruding 5′-termini and blunt ended fragments. There may, however, be difficulty in getting effective dephosphorylation where protruding 3′-termini shield the terminal 5′-phosphate (e.g., PstI fragments).
3. Phenol extraction and ethanol precipitation of small quantities of DNA may result in unacceptably low yields unless performed very carefully. Siliconized tubes can be used for the ethanol precipitation, which prevents the DNA sticking to the sides of the tube. Do NOT use tRNA as a carrier if you are precipitating prior to ligation; the presence of tRNA interferes with the action of the ligase. Glycogen can be used as an alternative carrier.
4. With phage lambda, the success of infection depends on religation of the *cos* site before or after the DNA enters the cell. Phosphatase treatment of digested linear lambda DNA will dephosphorylate not only the restriction-enzyme cleavage sites, but also the terminal cos sequences, thus rendering the DNA non-infective. It is possible (but clumsy) to use phosphatase with a lambda vector, e.g., by ligating the *cos* sites on the vector *before* cleaving the DNA with a restriction enzyme. The cut restriction sites can then be dephosphorylated and ligated with the insert DNA as described above. Fortunately, phosphatase treatment is not usually necessary with lambda cloning (particularly with replacement vectors) since the packaging constraints apply a direct selection for phage DNA molecules carrying an inserted DNA fragment.

## References


1. Seeburg, P. H., Shine, J., Marshall, J. A., Baxter, J. D., and Goodman, H. M. (1977) Nucleotide sequence and amplification in bacteria of structural gene for rat growth hormone. *Nature* **220,** 486–494.
2. Ullrich, A., Shine, J., Chirgwin, J., Pictet, R., Tischer, E., Rutter, W. J., and Goodman, H. M. (1977) Rat insulin

genes: construction of plasmids containing the coding sequences. *Science* **196,** 1313–1319.
3. Efstratiadis, A., Vournakis, J. N., Donis-Keller, H., Chaconas, G., Dougall, D. K., and Kafatos, F. C. (1977) End labeling of enzymically decapped mRNA. *Nucleic Acid Res.* **4,** 4165–4174.

# Chapter 34

# Bacterial Transformation

***Elliot B. Gingold***

*Division of Biological and Environmental Sciences, The Hatfield Polytechnic, Hatfield, Hertfordshire, England*

## Introduction

If one consults any genetics textbook written before the mid-1970s, one will find much on the transformation of such bacterial species as *Bacillus subtilis*, but no mention of *Escherichia coli*. The method now most generally used for introducing DNA into *E. coli* is based on a study by Mandel and Higa in 1970 (*1*), who demonstrated that calcium chloride treatment would greatly enhance the ability of the cells to take up naked bacteriophage λ DNA. Later it was shown that the same method would also allow uptake of other DNA species including, in particular, plasmid DNA.

The method itself is remarkably simple and can be performed in less than 3 h once a culture of rapidly growing log phase cells have been obtained. The first stage involves producing competent cells by treating with cold calcium chloride. In the second, the DNA is introduced with the help of a heat shock. In the final stage the bacteria are left in nonselective medium to allow recovery and plasmid establishment before plating onto selective media.

The strain we have used as a general host is JA221, which originates in the laboratory of Carbon. There are, however, many other strains in general use and some in fact give even better results. An important property is, however, that the host should lack its own restriction system so that incoming unmodified DNA is not immediately cut up. Apart from this, the choice of host strain will probably depend on the nature of the markers on the plasmid and the overall aim of the exercise.

## Materials

1. LB broth: Yeast extract 0.5%, NaCl 1%, tryptone 1%.
2. LB agar: As above, plus 2% agar prior to autoclaving.
3. 0.1*M* $CaCl_2$.

Antibiotics are added to the above media after autoclaving: tetracycline to a final concentration of 15 µg/mL and ampicillin to 50 µg/mL. Solutions of these antibiotics are prepared with ampicillin at 50 mg/mL in slightly alkaline distilled water and tetracycline at 15 mg/mL in ethanol.

## Method

1. Prepare a small, overnight culture of the bacteria in LB broth. Grow at 37°C without shaking.
2. About 2 h before you are ready to begin the main procedure, use 1.0 mL of the overnight culture to inoculate 100 mL of fresh LB broth. This culture is grown with rapid shaking at 37°C until it reaches roughly $5 \times 10^7$ cells/mL. This corresponds to an $OD_{650}$ of about 0.4 for our cultures, but you should calibrate this for each of your own strains.
3. Take a 5 mL aliquot for each transformation reaction and transfer to sterile plastic centrifuge tubes. Cool on ice for 10 min.

4. Pellet the cells by spinning for 5 min at 5000*g*. It is necessary for the centrifugations to be performed at 4°C. We have found a refrigerated bench centrifuge ideal for this.
5. Pour off the supernatant and resuspend cells in 2.5 mL of cold 0.1*M* $CaCl_2$. Leave on ice for at least 20 min.
6. Centrifuge as in Step 3. You should observe a more diffuse pellet than previously. This is an indication of competent cells.
7. Resuspend the cells in 0.2 mL of cold 0.1*M* $CaCl_2$.
8. Transfer the suspensions to sterile, thin-walled glass bottles or tubes. The use of glass makes the subsequent heat shocks more effective.
9. To each tube add up to 0.1 μg of DNA, made up in a standard DNA storage buffer such as TE to a volume of 100 μL. Leave on ice for 30 min.
10. Transfer to a 42°C waterbath for 2 min and return briefly to ice.
11. Transfer the contents of each tube to 2 mL of LB broth in a small flask. Incubate with shaking at 37°C for 60–90 min.
12. Plate 0.1 mL aliquots of undiluted, $10^{-1}$ and $10^{-2}$ dilutions onto LB plates to which the antibiotics to be used for selection have been added.
13. Incubate overnight at 37°C.

## Notes

1. This method generally gives $10^4$–$10^6$ transformants/μg of closed circle plasmid DNA. Do note that the relationship between amounts of DNA added and yield is not totally linear. Greater than 0.1 μg of plasmid DNA per tube will decrease transformation efficiency.
2. It is essential that the cells used are in a rapid growth phase when harvested. Do not let them approach stationary phase.
3. Cells can be stored at 4°C once competent. Holding cells in $CaC1_2$ at 4°C will in fact increase transformation efficiency although this declines with more than 24 h

storage. Long periods of storage can be achieved by freezing the competent cells.

4. The revival step is necessary both to allow plasmid establishment and to allow expression of the resistance genes.
5. One problem encountered on plating on ampicillin is that resistant colonies will often be surrounded by a region of secondary growth. This is caused by the β-lactamase activity of the resistant cells hydrolyzing the surrounding antibiotic and thus allowing surviving sensitive cells to begin to grow. This problem can be avoided by using freshly made ampicillin plates and removing plates from the incubator promptly after the period of overnight growth.

## References

1. Mandel, M., and Higa, A. (1970) Calcium-dependent bacteriophage DNA infection. *J. Mol. Biol.* **53,** 109–118.

# Chapter 35

# Bacterial Transformation (Kushner Method)

***J. W. Dale and P. J. Greenaway***

*Department of Microbiology, University of Surrey, Guildford, Surrey and Molecular Genetics Laboratory, PHLS Centre for Applied Microbiology and Research, Porton Down, Salisbury, Wilts., United Kingdom*

## Introduction

The principle of this procedure is very similar to that of the standard calcium chloride method, (*see* Chapter 34) i.e., exponential phase cells are harvested and washed with a solution containing divalent cations. This renders the cells *competent,* which simply means they are now able to take up DNA from the solution. After mixing the competent cells with the DNA, they are subjected to a heat shock to promote the uptake of DNA, presumably by affecting the physical state of the lipids in the membrane.

This method, introduced by Kushner (*1*), differs from the standard procedure in the use of rubidium chloride, buffered with MOPS (morpholinopropane sulfonic acid), rather than calcium chloride for washing the cells. The advantage is that the efficiency of transformation is considerably higher—up to $10^7$ tranformed cells being obtained per microgram of intact plasmid DNA.

## Materials

1. Strain: *E. coli* HB101. Many other strains are also suitable, although the frequency of transformation may vary quite considerably from one strain to another.
2. MOPS/RbCl: 10 m*M* MOPS, 10 m*M* RbCl, pH 7.0.
3. MOPS/$CaCl_2$/RbCl: 100 m*M* MOPS, 50 m*M* $CaCl_2$, 10 m*M* RbCl, pH 6.5.
4. L broth:

| | |
|---|---|
| Tryptone | 15 g |
| Yeast extract | 5 g |
| NaCl | 5 g |

   Adjust the pH to 7.2 with sodium hydroxide and make up to 1 L. Dispense in 100 mL amounts and sterilize by autoclaving.
5. Sterile 100 mL conical flasks.
6. Appropriate selection medium, i.e., L agar (solidified with 1% agar) supplemented with an antibiotic to select for the plasmid concerned.

## Method

1. Set up an overnight culture of HB101 in L broth.
2. Use 0.1 mL of the overnight culture to inoculate 20 mL of L broth in a 100 mL flask. (This will provide enough cells for 12 separate transformations). Grow the culture, with shaking, at 37°C until the $OD_{650}$ is about 0.15.
3. Dispense 1.5 mL aliquots of the culture in microfuge tubes and spin for 5 min in a microcentrifuge at 4°C. Alternatively, 15 mL corex tubes can be used in a refrigerated centrifuge, in which case the cells are pelleted at 8000*g* for 10 min.
4. Resuspend the cell pellets in 0.5 mL of ice-cold MOPS/RbCl. Keep the tubes on ice during this and subsequent steps.

5. Immediately recover the cells by centrifugation for 5 min at 4°C, and resuspend the cells in 0.5 mL of ice-cold MOPS/$CaCl_2$/RbCl buffer. Keep the cells on ice for 60 min.
6. Pellet the cells at 4°C and resuspend in 0.2 ml of ice-cold MOPS/$CaCl_2$/RbCl. The cells are now competent and can be stored on ice for several hours before use.
7. Add 10–20 μL of DNA (this should be less than 40 ng of DNA) to each tube of competent cells. Leave on ice for 30–45 min, with occasional mixing.
8. Transfer the tubes to a 45°C waterbath for 2 min.
9. Add 2 mL of L broth to each tube and incubate at 37°C for at least 1 h. This is to allow time for the bacteria to recover and to express the antibiotic resistance (or other) genes before being plated on the selective medium.
10. Make appropriate dilutions (in L broth) of each transformation mix: $10^{-1}$, $10^{-2}$, and $10^{-3}$ in the first instance (*see* Note 5).
11. Plate 0.1 mL aliquots of the diluted and undiluted transformation mixes onto appropriate selection plates. After allowing the plates to dry, incubate them overnight at 37°C. Store any remaining transformation mixes at 4°C overnight pending the initial plating results.

# Notes

1. Do not grow the culture too long before preparing competent cells. It is a fallacy to believe that a greater number of cells will give you a better transformation frequency. Low density cultures are optimal for transformation. Similarly, do not try to scale up the preparation of competent cells. The procedure has been found to be effective only with small volumes.
2. Remember that competent cells are fragile. Do not exceed the conditions stated above during centrifugation, and be gentle when resuspending the cell pellets.

3. The heat shock step is a critical one. Make sure the tube contents reach at least 42°C. Do not incubate too long at 45°C or else a loss in viability will result.
4. Competent cells may be stored for several months at −70°C by adding an equal volume of 40% glycerol at the end of Step 6. The competent cells must be recovered by centrifugation after storage and then resuspended in 0.1 mL of fresh MOPS/$CaCl_2$/RbCl before use.
5. The final cell suspensions may require dilution before plating out. However, better results are obtained by diluting the DNA before transformation, and doing replicate transformations.
6. If a large number of transformants are required, the whole of the transformation mixes should be plated out immediately rather than storing a portion overnight—the number of colonies obtained is usually less when the mixture is plated the following day.
7. Do not be tempted into trying to increase the number of transformants by increasing the amount of DNA used. A small increase in the amount of DNA will have little effect; a larger increase can result in a *reduction* in the number of transformants obtained because of competition for uptake. Once again, the correct procedure is to run replicate transformations.

## References

1. Kushner, S. R. (1978) An improved method for transformation of *Escherichia coli* with ColE1-derived plasmids. In *Genetic Engineering* (Boyer, H. B., and Nicosia, S. eds.) pp. 17–23. Elsevier/North Holland, Amsterdam.

# Chapter 36

# In Vitro Packaging of DNA

*J. W. Dale and*
*P. J. Greenaway*

*Department of Microbiology, University of Surrey, Guildford, Surrey and Molecular Genetics Laboratory, PHLS Centre for Applied Microbiology and Research, Porton Down, Salisbury, Wilts., United Kingdom*

## Introduction

In the normal growth cycle of bacteriophage lambda, the proteins that ultimately form the head of the phage particle are assembled into an empty precursor of the head (prehead); the phage DNA is replicated separately and then inserted into the empty head particles—a process known as packaging. A number of phage gene products play an important role in this process. Amongst these are:

(i) The E protein, which is the major component of the phage head; mutants that are defective in this gene are unable to assemble the preheads, and therefore accumulate the other, unassembled, components of the phage particle as well as the other proteins involved in packaging.

(ii) The D and A proteins, which are involved in the packaging process itself. Mutants that are defective in these genes are able to produce the preheads, but will not package DNA. This results in the accumulation of empty preheads.

In vitro packaging involves the use of two bacterial extracts: one from cells infected with a D mutant phage, and a second from cells infected with a phage mutant in gene E. Mixing these two extracts allows in vitro complementation and results in the ability to assemble mature phage particles. However, the packaging reaction will only work for DNA with certain properties. Lambda DNA is replicated as a multiple length linear molecule; the enzymes involved in packaging (in particular, the A protein) will recognize a specific site on this DNA (the *cos* site), cut the DNA at that point, and start packaging it into the phage head. Packaging continues until a second *cos* site is reached, when a second cut is made, giving a single complete lambda DNA molecule inside the phage head. (The cleavage at the *cos* sites is asymmetric, giving rise to the single-stranded cohesive ends of mature linear lambda DNA.) It therefore follows that only DNA molecules that carry a *cos* site will be packaged.

Furthermore, in order to produce phage particles that are capable of infecting a bacterial cell, the length of DNA between two *cos* sites must be within certain limits (packaging limits). If this length is greater than 105% of the size of lambda DNA (i.e., greater than about 53 kb), the head fills up before the second *cos* site is reached and the second cleavage cannot then take place. If the DNA is less than about 80% of the size of lambda DNA, cleavage will occur when the second *cos* site is reached, but the DNA within the phage head will then be insufficient to maintain the structural integrity of the particle.

The existence of packaging limits in one sense reduces the usefulness of lambda phage vectors, since it places constraints on the size of the inserts that can be cloned; however, it can be turned to advantage by permitting a selection to be applied for inserts within a desired size range.

A further application of in vitro packaging arises in the use of a special type of cloning vector known as a cosmid (*see* Chapter 49), which is a plasmid carrying the *cos* sequence. The packaging reaction depends on the presence of *cos* sites separated by a suitable length of DNA, but is not selective for the nature of the DNA between the these sites. The cosmid vector itself is much too small to be packaged into a viable particle, and thus a large fragment of foreign DNA must be inserted before a molecule in the correct size range for packaging is obtained.

In vitro packaging therefore has two advantages over transformation/transfection as a means of introducing DNA into a bacterial cell. First, it is more effective, especially when using large DNA molecules such as lambda vectors, or cosmids with large inserts. Second, it provides a selection for the presence of these large inserts.

It is important to realize that this introduction is intended as a brief introduction to a complicated subject. It is advisable to consult specialized texts before attempting to use lambda or cosmid vectors or in vitro packaging. The procedure described in this chapter includes the protocols for preparing the packaging mixes although these are now available commercially.

# Materials

1. Packaging strains:
   BHB2688: lysogenic for $CI_{ts}$ b2 red3 Eam4 Sam7
   BHB2690: lysogenic for $CI_{ts}$ b2 red3 Dam15 Sam7
   The $CI_{ts}$ mutation makes the lysogen heat inducible, because the mutant repressor is thermolabile. The b2 deletion and the red3 mutation reduce the background arising from the presence of endogenous DNA in the packaging extracts. The Sam7 mutation makes the phage lysis defective, thus allowing the phage products to accumulate within the cell after induction of the lysogen. The purpose of the D and E mutations is described in the introduction to this chapter. The D, E, and S mutations are amber mutations, thus allowing the phage to be grown normally within an amber suppressor host.

2. L broth:

| | |
|---|---|
| Tryptone | 15 g |
| Yeast extract | 5 g |
| NaCl | 5 g |

Dissolve the ingredients in about 600 mL of water, adjust the pH to 7.2 with NaOH, and make the volume up to 1 L. Transfer 500 mL to each of two 5-L flasks and sterilize by autoclaving.
3. 10% Sucrose in 50 m*M* Tris-HCl, pH 7.5.
4. Lysozyme (2 mg/mL in 0.25*M* Tris-HCl, pH 7.5), prepared fresh.
5. Liquid nitrogen.
6. Packaging buffer: 5 m*M* Tris-HCl, pH 7.5; 30 m*M* spermidine; 60 m*M* putrescene; 20 m*M* $MgCl_2$; 15 m*M* ATP; 3 m*M* 2-mercaptoethanol.
7. Sonication buffer: 20 m*M* Tris-HCl, pH 8.0; 3 m*M* $MgCl_2$; 5 m*M* 2-mercaptoethanol; 1 m*M* EDTA.
8. Phage buffer:

| | |
|---|---|
| $Na_2HPO_4$ | 7 g |
| $KH_2PO_4$ | 3 g |
| NaCl | 5 g |
| $MgSO_4 \cdot 7H_2O$ | 0.25 g |
| $CaCl_2$ | 0.015 g |
| gelatin | 0.01 g |

Add water to 1 L. Dispense in small volumes and sterilize by autoclaving.
9. Sensitive *E. coli* strain.

# Method

## *Preparation of Packaging Mix A (Packaging Protein)*

1. Set up an overnight culture of BHB2688 in L broth at 30°C.
2. Use the overnight culture to inoculate 500 mL of L broth and incubate, with shaking, at 30°C. When the $OD_{650}$ has reached 0.3, remove the flasks from the shaker and place in a 45°C waterbath for 15 min, without shaking. This is to induce the phage.

3. Transfer the flask to an orbital shaker and incubate at 37°C for 1 h.
4. Recover the cells by centrifugation at 9000*g* (8000 rpm in a 6 × 250 mL rotor) for 10 min.
5. Remove all the supernatant, invert the tubes, and allow them to drain for at least 5 min.
6. Resuspend the cells in approximately 2 mL of sucrose–Tris buffer. Distribute the suspension in 0.5 mL aliquots, and to each tube add 50 μL of a freshly prepared solution of lysozyme. Mix gently by inversion.
7. Freeze this suspension immediately in liquid nitrogen.
8. Remove the tubes from the liquid nitrogen (**NB:** use forceps!), and allow the cells to thaw at 0°C for at least 15 min.
9. Pool the thawed cells, add 50 μL of packaging buffer and centrifuge at 48,000*g* (20,000 rpm in an 8 × 50 rotor) for 30 min. Recover the supernatant and dispense 50 μL aliquots into cooled microcentrifuge tubes. Freeze the tubes in liquid nitrogen and then store at −70°C until required.

## *Preparation of Packaging Mix B (Phage Heads)*

1. Using strain BHB 2690, carry out steps 1–5 as above.
2. Resuspend the cells in 5 mL of cold sonication buffer. Sonicate the cells until the suspension is no longer viscous. Keep the vessel in an icewater bath during the process, use short bursts of sonication (3–5 s), and avoid foaming.
3. Centrifuge the suspension at 15,000*g* (10,000 rpm in a 6 × 250 mL rotor) for 10 min.
4. Recover the supernatant and dispense 50 μL aliquots into cooled microcentrifuge tubes. Freeze each tube immediately in liquid nitrogen and then store at −70°C until required.

## *Packaging reaction*

1. Thaw one tube of each extract (A and B) as prepared above.
2. Add 1–2 μL of the DNA to be packaged to 7 μL of sonication buffer in a microcentrifuge tube.

3. Add 1 μL of packaging buffer, 10 μL of extract A, 6 μL of extract B. Incubate at 25°C for 60 min.
4. Dilute with 0.5 mL of phage buffer and plate out using an appropriate lambda-sensitive host onto either BBL agar overlay plates (for lambda vectors) or onto antibiotic selection plates (for cosmids).
5. Score the resulting plaques or transformants and screen them for the presence of specific inserted DNA sequences, e.g., by an *in situ* hybridization procedure (Chapters 42 and 43).

## Note

1. There are several alternative procedures for the preparation of packaging extracts. Examples of other procedures will be found in the references listed below.

## References

1. Collins, J. and Hohn, E. (1978) Cosmids: A type of plasmid gene cloning vector that is packageable *in vitro* in bacteriophage heads. *Proc. Natl. Acad. Sci. USA* **75,** 4242–4246.
2. Enquist, L., and Sternberg, N. (1979) *In vitro* packaging of lambda Dam vectors and their use in cloning DNA fragments. *Meth. Enzymology* **68,** 281–298.
3. Hohn, B. (1979) In vitro packaging of lambda and cosmid DNA. *Meth. Enzymology* **68,** 299–309.
4. Hohn, B., and Murray, K. (1977) Packaging recombinant DNA molecules into bacteriophage particles *in vitro. Proc. Natl. Acad. Sci. USA* **74,** 3259–3263.
5. Sternberg, N., Tiemeier, D., and Enquist, L. (1977) In vitro packaging of a lambda dam vector containing EcoRI DNA fragments of *Escherichia coli* and phage P1. *Gene* **1,** 255–280.

# Chapter 37

# Yeast Transformation



## *Elliot B. Gingold*

*Division of Biological and Environmental Sciences, The Hatfield Polytechnic, Hatfield, Hertfordshire, England*


## Introduction

The yeast *Saccharomyces cerevisiae* has tremendous advantages as a host in gene cloning experiments. It is a microorganism for which most of the techniques developed in bacterial work can be applied, including chemical mutagenesis, selective plating, and replica plating. Being the basis of an ancient industry, its fermentation characteristics are well understood. It is also a eukaryote with mitotic and meiotic divisions, cellular compartmentalization, and post-translational modification of proteins similar to that seen in higher species. Yeast is thus an ideal organism to turn to in experiments for which bacterial hosts are not suitable, but microbial techniques are nonetheless required.

A number of plasmid systems have been used in yeast cloning. These can be grouped into integrating plasmids (YIp), which can only replicate by integrating into a chromosome, replicating plasmids (YRp), which include a yeast chromosomal origin of replication, and vector systems, which are based on the natural yeast 2μ plasmid (YEp). The YIp plasmids will give only a low frequency of transformation while the other classes will give up to $10^5$ transformant/μg DNA. Yeast cloning plasmids normally also contain all or part of an *Escherichia coli* plasmid. Such 'shuttle' vectors can thus be propagated in bacterial cells as well as yeast cells. This has major advantages since bacterial cloning can be used to provide far greater amounts of the plasmid DNA than would be possible directly from yeast cells.

To transform yeast cells it is necesary to first remove the cell wall and produce protoplasts (sometimes referred to, perhaps more accurately, as spheroplasts). This is generally done by weakening the cell wall with a reducing agent and then subjecting it to enzymic degradation. The procedures for getting DNA into cells are in fact very similar to those previously developed for protoplast fusion and involve the joint action of polyethylene glycol (PEG) and $Ca^{2+}$ ions. Protoplasts must then be regenerated by embedding them in osmotically buffered agar. In general, recombinants are selected on the basis of complementation of a nutrient deficiency of the host.

The method here closely follows that of Beggs (*1*). Some of the steps can be cut down, but this leads to less reliable results. In this Chapter, an experiment involving transformation of a histidine-requiring yeast (his 3) with a his $3^+$ plasmid is described. Variations involving other plasmid–host systems are easily made.

# Materials

1. Yeast — A 3617 C (a his 3 gal 2) (2)
2. Plasmid — YEp 6 ($amp^r$, his $3^+$) (2)
3. Enzyme — Novozym 234 (Novo, Copenhagen, Denmark)
4. Reducing buffer: — 1.2*M* sorbitol, 25 m*M* EDTA, 50 m*M* dithiothreitol, pH 8.
5. Enzyme buffer: — 1.2*M* sorbitol, 0.01*M* EDTA, 0.1*M* sodium citrate, pH 5.8
6. Sorbitol: — 1.2*M* sorbitol

7 Sorbitol–calcium chloride: 1.2*M* sorbitol, 10 m*M* $CaCl_2$
8. Fusion buffer: 20% polyethylene glycol 4000, 10 m*M* $CaCl_2$, 1 m*M* Tris, pH 7.5
9. Growth Media
   (*a*) *YEPD:* Yeast extract (1%), bacteriological peptone (2%) glucose (2%)
   (*b*) *Minimal:* 0.67% Difco yeast nitrogen base, 2% glucose, 1.2*M* sorbitol, 2% agar
   (*c*) *Supplemented:* As minimal plus 20 mg/L histidine

# Method

1. Grow the yeast in YEPD at 30°C overnight with shaking. It is preferable to aim for late log phase ($10^8$ cells/mL), but this is not essential.
2. Harvest 2 × 12 mL of cell suspension in sterile plastic centrifuge tubes. A short spin in a bench centrifuge should be adequate.
3. Wash the pellets in sterile distilled water.
4. Suspend the pellets in 4 mL of reducing buffer and incubate at 30°C for 10 min with gentle shaking.
5. Centrifuge and wash the pellets twice in 12 mL of 1.2*M* sorbitol. All centrifugation steps from this point on are performed with ice-cold reagents and in a refrigerated bench centrifuge or in the cold room.
6. Take the pellets up in 3 mL of enzyme buffer and add 1 mL of 20 mg/mL Novozym 234.
7. Incubate at 30°C for 1 h with gentle shaking. During this time, protoplasting can be followed microscopically by comparing cell counts in sorbitol and 1% SDS (protoplasts will lyse in SDS).
8. Centrifuge protoplasts and resuspend in 12 mL of cold 1.2*M* sorbitol. Wash three times in this buffer. Handle protoplasts gently! Do not agitate violently to resuspend or centrifuge with excess speed. 1000*g* for 10 min should be adequate.
9. Resuspend in 0.1 mL of sorbitol–calcium chloride. This should give around $10^{10}$ cells/mL. To one tube add plasmid DNA; the other is a control.

10. Add the DNA in a volume no greater than 20 μL. In general, around 1 μg of plasmid DNA is ideal, but smaller quantities are acceptable.
11. Leave at room temperature for 15 min, then add 1 mL of the fusion buffer.
12. After a further 15 min, centrifuge the cells down (gently as in previous steps).
13. Resuspend in 0.1 mL of sorbitol–calcium chloride, plus 0.05 mL of YEPD media including 1.2*M* sorbitol. Incubate for 20 min at 30°C. This 'revival' step is not needed for all strains.
14. Make the volume up to 2 mL with 1.2*M* sorbitol.
15. Have prepared in a 45°C water bath 10 mL aliquots of molten minimal and supplemented media. The 0.1 mL aliquots of the cell suspension (or dilutions in 1.2*M* sorbitol) are added to the molten agar, quickly mixed, and poured into a sterile empty petri dish. The minimal media reveals transformants, whereas the supplemented plates will measure regeneration efficiency. The $10^{-2}$, $10^{-1}$, and undiluted samples should be plated in minimal media and $10^{-6}$, $10^{-5}$, and $10^{-4}$ dilutions plated in supplemented media. Regeneration as well as transformation frequency can be variable.
16. Incubate plates at 30°C. Be patient, with some plasmids colonies may not appear for 5 d!
17. Colonies growing in minimal plates can be picked and transferred onto fresh minimal plates (Sorbitol is no longer needed). Transformed colonies will generally be unstable and could lose the plasmid if grown on complete (e.g., YEPD) media.

## Notes

1. In the procedure the enzyme Novozym 234 has been used. We have found this to be cheap, efficient, and relatively sterile. Alternatively, one can use β-glucanases such as β-glucuronidase (Sigma), helicase (Industrie Biologique Francaise, France), or zymolase (Kirin Brewery, Tokyo, Japan). Some preparations need filter sterilization and this can be difficult because of their viscous nature.

2. Different strains will have differing susceptabilities to protoplasting. If protoplasting appears complete well before 1 hr, proceed. Excessive protoplasting often leads to poor regeneration. Other strains may of course require longer than 1 h. You will have to establish this for your system.
3. It is difficult to measure regeneration frequencies precisely because of the clumping effect of the PEG treatment and uncertainty about the degree of break up of these clumps in subsequent dilutions. Nonetheless, either a high ( > 30%) or low ( < 1%) regeneration is not usually accompanied by good transformation frequencies.
4. Use a strain known to give good transformation frequency, experiment with others only when you feel confident of your technique. Many yeast strains give very poor transformation results. If you want to get a plasmid into a strain with a particular genotype, some traditional genetic manipulation to combine your desired characteristics with good transformation ability will probably be needed!
5. Unfortunately, different batches of PEG will give differing results. It is well worth trying different quantities of PEG in the fusion buffer to optimize conditions.
6. As a cheaper alternative to Difco nitrogen base, it is entirely satisfactory to use laboratory made minimal media, such as that of Wickerham (1946).

# References


1. Beggs, J. D. (1978) Transformation of yeast by a replicating hybrid plasmid. *Nature* **275,** 104–108.
2. Struhl, K., Stinchcomb, D. T., Scherer, S., and Davis, R. W. (1979) High-frequency transformation of yeast: Autonomous replication of hybrid DNA molecules. *Proc. Natl. Acad. Sci. USA* **76,** 1035–1039.
3. Wickerham, L. J. (1946) A critical evaluation of the nitrogen assimilation tests commonly used in the classification of yeast. *J. Bacteriol.* **52,** 293–301.

# Chapter 38

# Radiolabeling of DNA by Nick Translation


## C. G. P. Mathew

*MRC Molecular and Cellular Cardiology Research Unit, University of Stellenbosch Medical School, Tygerberg, South Africa*


## Introduction

Nick translation is the name given to a reaction that is used to replace cold nucleoside triphosphates in a double-stranded DNA molecule with radioactive ones (*1*,*2*). Free 3′-hydroxyl groups are created within the unlabeled DNA (nicks) by deoxyribonuclease 1 (DNAse 1). DNA polymerase 1 from *E. coli* will then catalyze the addition of a nucleotide residue to the 3′-hydroxyl terminus of the nick. At the same time, the 5′- to 3′-exonuclease activity of this enzyme will eliminate the nucleotide unit from the 5′-phosphoryl terminus of the nick. Thus a new nucleotide with a free 3′-OH group will have been incorporated at the position where the original nucleotide was excised, and the nick will have been shifted along by one nucleotide unit in a 3′ direction. This 3′ shift, or translation, of the nick will result in the sequential addition of new nucleotides to the DNA while the pre-existing nucleotides will be removed. If radioactively labeled deoxyribonucleoside triphosphates are used as substrates, up to 50% of the residues in the DNA can be labeled. Furthermore, Rigby et al. have shown (2) that the DNA is labeled throughout at a uniform specific activity, which is an important requirement if the DNA is to be used as a probe in molecular hybridization experiments.

Once the DNA has been labeled, it can be used to detect complementary sequences on Southern blots (Chapter 9), or in recombinant organisms by colony or plaque hybridization (Chapters 42 and 43). Nick translation also provides a means of detecting picogram amounts of DNA on a gel by autoradiography, whereas ethidium bromide staining will only detect about 50–100 ng of DNA per band.

Further details of the reaction mechanism and its application can be found in ref. (3).

## Materials

1. Since the procedure usually involves the use of high energy emitters of high specific activity, a transparent perspex shield should be constructed to protect workers from prolonged exposure. The perspex should be 0.5–1.0 cm thick. Convenient dimensions of the shield are 50 cm (height) × 30 cm (width). The laboratory should also be equipped with a portable Geiger counter for monitoring clothing, benches, and so on.
2. 10 × Nick translation buffer: 0.5*M* Tris-HCl (pH 7.8); 50 m*M* $MgCl_2$; 100 m*M* 2-mercaptoethanol; 100 μg/mL nuclease-free bovine serum albumin.
3. 10 × Stock solutions of deoxynucleoside triphosphates: 0.2 m*M* of each of dCTP, dATP, dGTP, and dTTP.
4. *E. coli* DNA polymerase 1: 0.5–1.0 U/μL.
5. DNase 1 (e.g., pancreatic DNase from Sigma), 50–100 pg/μL. Dilute in 10 m*M* Tris-HCl (pH 7.5), 5 m*M* $MgCl_2$ just before use.

6. Labeled deoxynucleoside triphosphate (*see* Note 2), e.g., [α-$^{32}$P] dCTP, 10 mCi/mL, 3000 Ci mmol (aqueous solution).
7. Substrate DNA: 0.5 μg at a concentration of 0.02–0.5 mg/mL (*see* Note 3).
8. Stop buffer: 10 m*M* EDTA (pH 7.5), 0.1% SDS.
9. Sterile distilled $H_20$.
10. Phenol–chloroform: Melt redistilled phenol and add to an equal volume of chloroform. Shake with 0.5 vol × TE (*see* 11), and stand until phases have separated.
11. 1 × TE buffer: 10 m*M* Tris-HCl (pH 7.5), 1 m*M* EDTA.
12. Sephadex G-50 (medium, equilibrated in 1 × TE).

All the reagents for the nick translation reaction should be stored at −20°C. Phenol–chloroform (in a light-tight bottle) and Sephadex G-50 slurry are stored at 4°C. $^{32}$P has a half-life of 14.3 d, so that labeled nucleotides should be ordered only shortly before they are required.

# Method

1. Pipet the following into a sterile 1.5 mL microfuge tube:

   5 μL 10 × nick translation buffer
   5 μL each of 0.2 m*M* dGTP, dATP, and dTTP
   10 μL [α-$^{32}$P] dCTP (*see* Note 4)
   2 U DNA polymerase 1
   200 pg DNase 1 (*see* Note 5)

   Make up to 50 μL with sterile distilled $H_20$ and incubate for 60 min at 15°C (*see* Note 6).
2. Add 100 μL of stop buffer, and extract with an equal volume of phenol–chloroform. Pipet the aqueous phase into a clean microfuge tube, and re-extract the organic phase with 100 μL TE. Pool the aqueous phases.
3. Prepare a Sephadex G-50 column in a Pasteur pipet and equilibrate with TE. Load the pooled aqueous phases onto the column and elute with TE. Collect 15 × 3 drop fractions (±120 μL).

4. Count 1.2 µL aliquots of the fractions. Pool fractions from the excluded peak and calculate the specific activity of the DNA.

$$\text{Specific activity} = \text{Total counts (excluded peak)} \times 100 \times 2 \times \frac{100}{\text{\% Counting efficiency}} \text{ dpm/}\mu\text{g}$$

Measure the volume of the excluded peak and count an aliquot of this if an accurate figure for specific activity is required.

## Notes

1. The reagents required for the nick translation are commercially available in kit form (e.g., Bethesda Research Laboratories or Amersham International). Nick translation buffer and the 3′ unlabeled nucleotides can be combined into one solution to reduce the number of pipeting steps. This can also be done with DNase 1 and DNA polymerase 1. The combined enzyme solution should be stored in 50% glycerol.
2. Nucleotides labeled with $^{3}H$, $^{32}P$, or $^{35}S$ may be used, as required. $^{32}P$ is normally used for Southern blots and colony or plaque hybridization. Nucleotides with a lower specific activity will, of course, produce probes of lower specific activity. Any, or all, of the four dNTPs can be used to label the DNA, but [α-$^{32}$P]-dCTP and -dTTP have been reported to be more stable than dATP and dGTP (4). A variety of labeled nucleotides are commercially available.
3. The amount of labeled DNA needed will depend on the experiment for which it is to be used. A 0.5 µg amount of probe is adequate for Southern blots. The reaction may be scaled up or down as required.
4. The addition of 100 µCi of [α-$^{32}$P]-dCTP at a specific activity of 3000 Ci/mmol to the reaction described in the Method section should produce a probe with a specific activity of 2–4 × $10^{8}$ dpm/µg. The amount of labeled nucleotide required to produce a probe of a particular specific activity can be calculated from a normogram (4).

5. The activity of commercial preparations of DNAse 1 is variable. A high DNAse 1 concentration will produce a probe of high specific activity, but reduced single-strand length.
6. Aliquots may be removed from the reaction at various times and the incorporation of label into the probe determined.
7. Poor incorporation of labeled nucleotide into the probe may be caused by several factors, e.g., chemical degradation of the labeled nucleotide, reduced activity of the enzyme preparations, and impurities in substrate DNA. The DNA may be further purified using a disposable ion-exchange column (Schleicher & Schuell Elutip).

# References

1. Maniatis, T., Jeffrey, A., and Kleid, D. G. (1975) Nucleotide sequence of the rightward operator of the phage λ. *Proc. Natl. Acad. Sci. USA* **72,** 1184–1188.
2. Rigby, P. W. J., Dieckmann, M., Rhodes, C., and Berg, P. (1977) Labelling DNA to high specific activity in vitro by nick translation with DNA polymerase 1. *J. Mol. Biol.* **113,** 237–251.
3. Mathew, C. G. P. (1983) Labelling DNA in vitro—Nick translation, in *Techniques in Molecular Biology* (Walker, J. M., and Gaastra, W., eds.) pp. 159–166. Croom Helm, London and Canberra.
4. The Radiochemical Centre, Amersham (1980) Labelling of DNA with $^{32}P$ by nick translation. Technical Bulletin 80/3.

# Chapter 39

# Radiolabeling of DNA Using Polynucleotide Kinase

***Wim Gaastra and Jytte Josephsen***

*Department of Microbiology, The Technical University of Denmark, Lyngby, Denmark*

## Introduction

Labeling of DNA fragments at the 5′ end with polynucleotide kinase is one of the methods currently employed to obtain highly labeled DNA for sequencing purposes. This end-labeled DNA is used in the partial chemical degradation method for sequencing DNA (*see* Chapter 51 and ref. 1).

The method involves the generation of DNA fragments with a so-called 5′ "sticky end." Therefore the restriction enzyme to be used should cleave in a similar way, for example, to *Bam*Hl and generate extensions at the 5′ end of the DNA (Fig. 1).

The phosphate group at the 5′ end of the DNA is removed by treating the DNA with alkaline phosphatase. The enzyme polynucleotide kinase is able to attach a labeled phosphate molecule to the 5′-OH group of the DNA

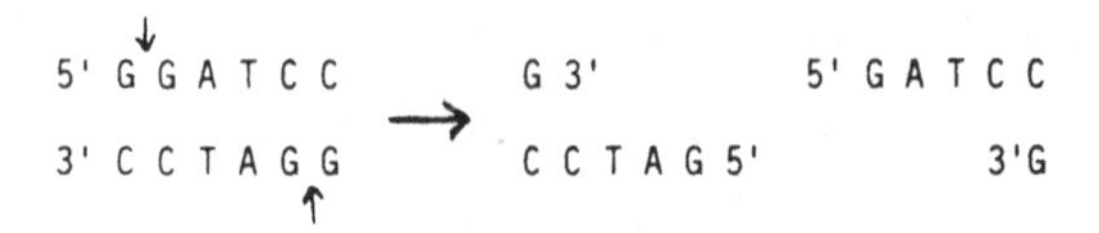


Fig. 1. Cleavage of a DNA fragment with *Bam*Hl to give 5′ extensions.

when the incubation is performed in the presence of γ-$^{32}$P-labeled ATP. The DNA is then cleaved with a second restriction enzyme that should give fragments of unequal size. The fragments are separated on the basis of their size differences, are isolated, and are then ready for sequencing.

## Materials

1. Alkaline phosphatase, either bacterial or calf intestinal (*see* Note 2).
2. Phosphatase buffer: 50 m*M* Tris-HCl, pH 8.0; 5 m*M* EDTA.
3. Dialysis buffer: 0.1*M* Tris-HCl, pH 7.9; 1 m*M* $MgCl_2$.
4. TE buffer: 10 m*M* Tris-HCl, pH 8.0; 1 m*M* EDTA.
5. Distilled phenol, saturated with TE buffer.
6. Diethyl ether saturated with $H_2O$.
7. 3*M* sodium acetate, pH 5.6; 10 m*M* $MgCl_2$.
8. Ethanol, 96%.
9. Kinase buffer (10 × ): 0.5*M* Tris-HCl, pH 7.6; 1 m*M* EDTA; 0.1*M* $MgCl_2$; 1 m*M* spermidine; 50 m*M* dithiothreitol.
10. Ammonium acetate, 2.5*M*.
11. tRNA solution, 1 mg/mL in TE buffer.
12. γ$^{32}$P-labeled ATP, of the highest specific activity available.
13. Imidazole-HCl buffer, pH 6.6 (required for alternative procedure only, *see* Note 5): 100 m*M* imidazole; 10 m*M* $MgCl_2$; 5 m*M* dithiothreitol; 0.1 m*M* spermidine; 0.1 m*M* EDTA; 0.5 m*M* ADP.

# Method

1. Dissolve the DNA (100 pmol of 5′ end) to be labeled in 150 μL of phosphatase buffer and add 4–8 units of alkaline phosphatase.
2. Incubate for 1 h at 65°C if using bacterial alkaline phosphatase (BAP) or 1 h at 37°C if using calf intestinal alkaline phosphatase (CIAP).
3. Inactivate the CIAP by heating at 65°C for 10 min (BAP cannot be heat activated).
4. Add 100 μL of distilled phenol; vortex for 30 s, and then centrifuge for 2 min.
5. Pipet the upper DNA-containing phase into a new Eppendorf tube, and repeat steps 4 and 5 until a total of five phenol extractions have been performed.
6. Remove any remaining phenol by extracting the DNA solution with 300 μL of diethyl ether in the same way as described for the phenol extraction (step 4). The upper phase in this case is the ether and the lower phase is the DNA solution. The ether extraction should also be done five times. Finally remove any remaining ether by heating for 15 min at 65°C.
7. Add 15 μL of 3*M* sodium acetate and 450 μL 96% ethanol. Precipitate the DNA after mixing by leaving at −70°C for 5 min or −20°C for 20 min.
8. Dry the pellet after spinning for 15 min in an Eppendorf centrifuge.
9. Vacuum dry the γ-$^{32}$P-ATP solution if it is an alcohol solution. Use a 1–4 times molar excess of ATP over 5′ ends. Do not use less than 20 pmol of ATP.
10. Dissolve the dry DNA pellet in 20 μL of distilled water and add 5 μL of 10 × kinase buffer and the desired amount of $^{32}$P-ATP solution. Usually this is 5–10 μL of the aqueous solution, depending on the specific activity. Alternatively, dissolve the vacuum dried γ-$^{32}$P-ATP in 10 μL of distilled $H_2O$ and add this to the mixture.
11. Add distilled $H_2O$ to make the volume up to 49 μL and then add 1 μL (~ 5 U/μL) of $T_4$ polynucleotide kinase. Mix and incubate for 30 min at 37°C.

12. Add 200 μL of 2.5*M* ammonium acetate, 1 mL of cold 96% ethanol, mix, and precipitate the DNA for 10 min at −70°C.
13. Centrifuge for 10 min in an Eppendorf centrifuge and discard the supernatant (very radioactive).
14. Dissolve the pellet in 300 μL of 0.3*M* sodium acetate and precipitate with 900 μL of cold ethanol (10 min at −70°C). Centrifuge, discard the supernatant, and then repeat this step once more.
15. Finally, dissolve the pellet in the appropriate buffer for cleavage with the second restriction enzyme (*see* Chapter 31).

# Notes

1. The labeling of DNA with polynucleotide kinase is very sensitive to the presence of certain compounds. Agarose for instance is known to contain compounds (sulfate) that inhibit the kinase reaction. DNA fragments for kinase labeling should therefore preferentially not be eluted from agarose gels, but from acrylamide gels. However, fragments eluted from agarose can be "cleaned up" by gel filtration on a small Sephadex G 150 column.
2. Bacterial alkaline phosphatase (BAP) is usually sold as an ammonium sulfate suspension. Since this sulfate can also inhibit the kinase reaction, it has to be removed by dialysis. For this purpose an amount of BAP solution containing 1 mg of BAP is pelleted in an Eppendorf tube. The supernatant is removed and the pellet dissolved in 100 μL of dialysis buffer. The lid is removed from the tube and dialysis tubing is fixed over the tube. The tube is then hung with the membrane down into dialysis buffer (1 L). After dialysis the BAP solution is made up to 1 mL. This stock solution (1 μg/μL) is kept at 4°C.
3. The presence of BAP also interferes with the kinase reaction since it removes the radioactive phosphate group. BAP is a very resistant enzyme and the various phenol extractions should therefore be performed very

carefully. It is better to leave some of the aqueous layer than to take some of the interface containing the denatured phosphatase. The phosphatase should also be removed carefully since it can stick to the ends of the DNA and thereby inhibit the kinase reaction.

4. If required, the DNA can also be obtained in a pure form after the kinase labeling by running the sample on a Sephadex G 150 column. The DNA containing fractions are pooled and lyophilized.
5. Since RNA is much more readily labeled with polynucleotide kinase than DNA, the presence of RNA molecules should be avoided.
6. Incorporation of $^{32}P$ into DNA can be checked by chromatography on Polygram Cel 300 PEI using 0.75*M* sodium dihydrogen phosphate, pH 3.5, as solvent. Load 1 μL of labeled DNA solution at the origin and following chromatography carry out autoradiography on the plate. Free phosphate will run at the solvent front and the free triphosphate nucleotides will run to the central region of the chromatogram. DNA will remain at the origin. The intensity of the autoradiogram at the origin is therefore a direct measure of the amount of incorporation of $^{32}P$ into DNA.
7. As mentioned in the introduction, polynucleotide kinase catalyses the transfer of the γ phosphate of ATP onto the 5′-hydroxyl ends of DNA fragments. This reaction, called the direct phosphorylation reaction, is possible when the original phosphate group has been removed by treatment with alkaline phosphatase.
8. Polynucleotide kinase is, however, also able to catalyze the exchange of the γ-phosphate group of ATP for the existing 5′-terminal phosphate (2). This reaction has to be carried out in an excess ATP and ADP in order to be efficient. The procedure is as follows:

    1. Resuspend the DNA in 50 μL of imidazole buffer.
    2. Add 1–5 μL of γ-$^{32}P$-ATP (high specific activity). Preferably more than 100 pmol of ATP should be added.
    3. Add 20 units of $T_4$ polynucleotide kinase.
    4. Incubate for 15–60 min at 37°C.
    5. Continue from step 12 of the main protocol.

## References


1. Maxam, A. M., and Gilbert, W. (1977) A new method for sequencing DNA. *Proc. Natl. Acad. Sci. USA* **74, 560–564.**
2. Maxam, A. M., and Gilbert, W. (1980) Sequencing end-labeled DNA with base-specific chemical cleavages. *Meth. Enzymol.* **65,** 499–560.

# Chapter 40

# Radiolabeling of DNA with 3′ Terminal Transferase

***Wim Gaastra and Per Klemm***

*Department of Microbiology, The Technical University of Denmark, Lyngby, Denmark*

## Introduction

In contrast to the end labeling methods described in chapters 39 and 41, end labeling with terminal transferase can be used for DNA fragments with 3′ protruding ends, 3′ recessed ends, and blunt ends. Terminal deoxynucleotidyl transferase (terminal transferase) adds ribonucleotides onto the 3′ ends of DNA fragments. It normally requires a single stranded DNA molecule as primer, but in the presence of cobalt ions ($Mg^{2+}$ being the normal cofactor) it will also accept double-stranded DNA. The enzyme uses ribonucleotide triphosphates as the precursors and the reaction normally produces 3′ tails on the DNA fragments (*1*). Alkaline hydrolysis is then required to produce a DNA fragment with a uniquely labeled 3′ end. Terminal transferase also catalyses the addition of 3′-deoxyadenosine-5′-triphosphate (cordycepin-5′-triphosphate) onto the ends of DNA fragments. This compound is a chain terminator; it does not contain a hydroxyl group in the 3′ position, and therefore forms only mono-addition products. This produces the uniquely labeled DNA fragments required in the Maxam and Gilbert method for sequencing DNA, with no need for other procedures to produce unique 3′-labeled ends.

## Materials

1. Terminal deoxynucleotidyl transferase (~20 U/μL).
2. Cacodylate buffer: 1*M* sodium cacodylate, pH 7.6; 0.25*M* Tris, pH 7.6; 2 m*M* dithiothreitol.
3. $CoCl_2$ solution, 10 m*M*.
4. α-$^{32}$P-3′-deoxyadenosine-5′-triphosphate (cordycepin-5′-triphosphate), 6000–8000 Ci/mmol.
5. 3*M* sodium acetate, pH 5.6; 10 m*M* $MgCl_2$. (Also used as a 1:10 dilution to provide 0.3*M* sodium acetate.)
6. Ethanol, 96%.

## Method

1. Dissolve the DNA to be labeled, to a maximum of 10 pmol of 3′ ends, in 15 μL of distilled $H_2O$.
2. Add 10 μL of Cordycepin-5′-triphosphate (α-$^{32}$P-labeled, at least 15–20 pmol), 3.5 μL of $CoCl_2$ solution, 3.5 μL of cacodylate buffer, and 2–3 μL of terminal transferase solution (should contain 30–40 U).
3. Incubate at 37°C for 30–60 min, then add 215 μL of $H_2O$, 25 μL of 3*M* sodium acetate, and 750 μL of 96% ethanol.
4. Mix, and precipitate the DNA for 10 min at −70°C or 30 min at −20°C.
5. Centrifuge in an Eppendorf centrifuge for 10 min then discard the supernatant (very radioactive).
6. Dissolve the pellet in 300 μL of 0.3*M* sodium acetate and precipitate with 900 μL of cold 96% ethanol (10 min at −70°C for 30 min at −20°C).
7. Repeat steps 5 and 6.
8. Dissolve the pellet in the appropriate buffer for cleavage with a second restriction enzyme or analyze the DNA directly on a gel.

## Notes

1. Although the 3′ end labeling with terminal transferase can be used on all kinds of restriction enzyme fragments, it is only used in our laboratory when other methods of labeling are unsuitable. The main reason for this is that the incorporation of radioactivity into the DNA is not nearly as high as the methods described in Chapters 39 and 41.
2. Like the labeling of DNA with polynucleotide kinase, DNA labeling with terminal transferase is also sensitive to contaminating salts and protein. It is therefore recommended that the DNA should be cleaned up by repeated ethanol precipitation before use.
3. The cacodylate buffer and $CoCl_2$ should be stored at −20°C and can then be kept for several months. The solutions should not become turbid. If they do, they should be replaced.
4. The efficiency of labeling by this method is in the order: 3′ protruding ends > blunt ends > 3′ recessed ends. This order is probably a result of increasing steric hindrance of the enzyme.
5. The incorporation of $^{32}P$ into DNA can be checked by the chromatographic method described in chapter 39 (Note 6).

## References

1. Roychoudbury, R., and Wu, R. (1980) Terminal transferase-catalyzed addition of nucleotides to the 3′ termini of DNA. *Meth. Enzymol.* **65,** 43–62.

# Chapter 41

# Radiolabeling of DNA with the Klenow Fragment of DNA Polymerase

***Wim Gaastra and Jytte Josephsen***

*Department of Microbiology, The Technical University of Denmark, Lyngby, Denmark*

## Introduction

The Klenow fragment of DNA polymerase I is a proteolytic fragment obtained by the treatment of DNA polymerase I with subtilisin. The fragment still has the polymerase and 3′–5′ exonuclease activity, but lacks the 5′–3′ exonuclease activity of the holoenzyme (*1*). Radiolabeling of DNA fragments is most efficient with DNA fragments that contain recessed 3′ ends (*see* Fig. 1, Chapter 39). The method described here is for generating 3′ end-labeled DNA fragments that are suited for sequencing by the method of Maxam and Gilbert (Chapter 51), but can in principle also be used when radioactively labeled DNA is required for other purposes (e.g., Southern hybridization). The principle of the method is shown

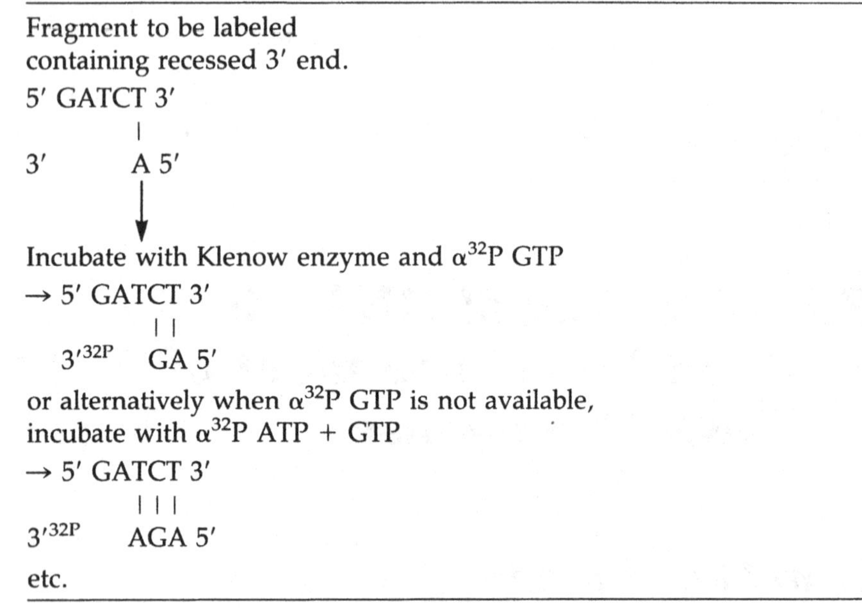


Fig. 1. Labeling of DNA with Klenow polymerase.

in Fig. 1. The DNA fragment to be labeled is incubated in the presence of one or more deoxyribonucleotide triphosphates (one of which is labeled with $^{32}$P in the α-phosphate group) and the Klenow fragment of DNA polymerase. As in the case with the kinase labeling of DNA (Chapter 39), the DNA fragment is cleaved with a second restriction enzyme, to generate fragments of unequal size that are labeled only in one end. The fragments are subsequently separated on the basis of their different molecular weights and isolated (Chapters 7 and 10).

## Materials

1. Klenow fragment of DNA polymerase.
2. Polymerase buffer ( × 10): 200 m*M* Tris-HCl, pH 7.6; 100 m*M* $MgCl_2$; 15 m*M* β-mercaptoethanol; 25 m*M* dithiothreitol.
3. If needed, cold dioxytriphosphates: 10 m*M* dTTP; 10 m*M* dATP; 10 m*M* dCTP; 10 m*M* dGTP.
4. α-$^{32}$P-labeled dioxytriphosphate.
   α-$^{32}$P dATP, α-$^{32}$P dCTP, α-$^{32}$P dTTP, or α-$^{32}$P dGTP,

(depending on which restriction site is to be labeled), of highest specific activity available.
5. TE buffer: 10 m*M* Tris-HCl, pH 8.0; 1 m*M* EDTA.
6. Distilled phenol, saturated with TE buffer.
7. Diethyl ether saturated with $H_2O$.
8. 3*M* sodium acetate, pH 5.6; 10 m*M* $MgCl_2$.
9. Ethanol, 96%.

# Method

1. Dissolve the DNA to be labeled (1–1000 ng) in 84 μL of $H_2O$ in an Eppendorf tube.
2. Add 10 μL of 10× polymerase buffer, 5 μL of $\alpha$-$^{32}$P-labeled dioxytriphosphate (50 μCi), and 1 μL of the Klenow fragment of DNA polymerase.
3. Incubate for 1 h at 25°C.
4. If considered desirable, add 10 μL of the cold dioxytriphosphate and incubate for a further 15 min.
5. Add 100 μL of distilled phenol, vortex for 30 s, then centrifuge for 2 min.
6. Transfer the upper DNA containing phase to a new Eppendorf tube, and remove any remaining phenol by extracting the DNA solution with 300 μ of diethyl ether in the same way as described for the phenol extraction. (**N.B.**: The upper phase in this case is the ether and the lower phase is the DNA solution.) The ether extraction should be repeated twice.
7. Add 10 μL of 3*M* sodium acetate and 300 μL of 96% ethanol, mix, and precipitate the DNA for 5 min at −70°C or 20 min at −20°C.
8. Centrifuge for 15 min in an Eppendorf centrifuge, remove the supernatant, and dry the pellet.
9. Dissolve the DNA in the correct buffer if another cleavage with a restriction enzyme is to be performed, or in 20 μL of $H_2O$ if the DNA is to be sequenced.

# Notes

1. The labeling of DNA with the Klenow fragment of DNA polymerase is not as sensitive to inhibition by certain compounds as is the kinase labeling (Chapter 39).

The procedure can therefore be used directly on fragments obtained by restriction enzyme digestion, and on fragments eluted from agarose gels.

2. $T_4$ DNA polymerase, which also lacks a 5′–3′ exonuclease activity, can be used instead of the Klenow fragment of DNA polymerase (2).
3. The cleavage sites generated by restriction enzymes like EcoR1 are not very suited for labeling with the Klenow fragment of DNA polymerase, because in this case there are two adjacent nucleotides that are the same

   (5′ AATTC 3′
   3′        G 5′).

   Since filling in with Klenow polymerase never proceeds to 100%, this gives rise to a mixture of labeled DNA fragments that differ by one nucleotide in size.

   5′ AATTC 3′ and 5′ AATTC 3′
   3′ $^{32P}$ AG 5′       3′ $^{32P}$A$^{32P}$AG 5′

   This size difference interferes with the reading of the sequence gels.
4. The incorporation of $^{32}P$ into DNA can be checked by the chromatographic method described in Chapter 39 (Note 6).
5. Depending on which $^{32}P$ dioxytriphosphates are available, it may be necessary to carry out the reaction in the presence of one or more cold dioxytriphosphates as well (*see*, for example, the lower part of Fig. 1). If this is the case, carry out the following step, and then continue at step 2 in the method section.

   (a) To the DNA add 10 μL of the cold 10 m*M* dNTP that is needed and then add water to make the volume up to 84 μL.

## References

1. Klenow, H., Overgaard-Hansen, K., and Patkar, S. A. (1971) Proteolytic cleavage of native DNA polymerase into two different catalytic fragments. *Eur. J. Biochem.* **22,** 371–381.
2. Challberg, M. D., and Englund, P. T. (1980) Specific labelling of 3′ termini with T 4 DNA polymerase. *Meth. Enzymol.* **65,** 39–43.

# Chapter 42

# Identification of Recombinant Plasmids by *In Situ* Colony Hybridization

***J. W. Dale and P. J. Greenaway***

*Department of Microbiology, University of Surrey, Guildford, Surrey and Molecular Genetics Laboratory, PHLS Centre for Applied Microbiology and Research, Porton Down, Salisbury, Wilts., United Kingdom*

## Introduction

After transformation, cells containing parental or recombinant plasmids are usually selected by taking advantage of an antibiotic resistance marker present on the vector plasmid. Cells containing recombinant plasmids can often be identified as containing recombinant plasmids by screening for the insertional inactivation of a second genetic marker on the plasmid. The presence of an insert can be verified, and its size determined, by restriction analysis of plasmid DNA isolated using a small-scale procedure such as that described in Chapter 27.

However, this does not provide any information on the nature of the inserted DNA. A method is therefore required that will not only confirm that recombinant plasmids have been obtained, but will also identify those recombinant colonies that contain the specific DNA sequence (gene) desired. This is usually done by *in situ* colony hybridization, a procedure originally published by Grunstein and Hogness (*1*). This procedure uses a radioactively labeled DNA fragment as a "probe" to test large numbers of bacterial colonies (e.g., from a "gene library") for the presence of a specific gene or fragments of a gene.

The procedure can be broken down into four parts:

(i) A suitable DNA fragment is radiolabeled by nick translation (this procedure is described in Chapter 38). This of course presupposes that it is possible to isolate the gene of interest or a related nucleotide sequence.

(ii) Bacterial colonies are transferred to a nitrocellulose membrane and lysed to release the DNA; this is then fixed to the filter by baking.

(iii) The filter with its bound pattern of DNA spots is incubated with the labeled probe; subsequently the filter is washed to release any non-specifically bound probe. The conditions of hybridization and subsequent washing can be adjusted to suit the particular circumstances, such as the $T_m$, and the degree of homology between the DNAs involved. In general terms it is best to hybridize at approximately 20–30°C below the $T_m$ of the DNA. The stringency of hybridization is lowered by increasing the salt concentration or by reducing the temperature. A 1% increase in formamide concentration reduces the melting temperature by 0.7°C. Because the kinetics of hybridization are concentration-dependent, it is better that these reactions are performed in the smallest volumes possible.

(iv) The filter is then subjected to autoradiography in a light-proof cassette using a sheet of X-ray film.

# Materials

## *For DNA Binding to Filters*

1. Nitrocellulose filter, of a size slightly smaller than the Petri dishes being used (i.e., 82 mm for a normal Petri dish). For sterilization, the filters must first be wetted by floating on the surface of a bath of water, then sandwiched between dry filter papers and wrapped in foil or placed in a glass petri dish for autoclaving. DO NOT expose nitrocellulose to dry heat, except in a vacuum oven.
2. Sodium hydroxide, 0.5*M*.
3. 1*M* Tris-HCl, pH 7.4.
4. 1.5*M* NaCl in 0.5*M* Tris-HCl, pH 7.4.
5. Whatman 3MM filter paper.
6. Equipment: vacuum oven.

## *For Hybridization*

7. Double strength SSC buffer (2 × SSC): 0.3*M* NaCl, 0.03*M* sodium citrate. Adjust the pH to 7.0 with sodium hydroxide. This buffer is usually made up as 20 × SSC (i.e., ten times the above concentrations), autoclaved, stored in aliquots at 4°C, and diluted just before use.
8. Formamide wash solution: 50% formamide (v/v) in 2 × SSC. The stringency of hybridization can be reduced by lowering the formamide concentration. **CAUTION!** Handle formamide with care.
9. 1*M* sodium hydroxide.
10. 1*M* HCl.
11. Equipment: plastic bag sealer.

## *For Autoradiography*

12. Whatman 3MM filter paper.
13. Kodak XRP5 X-ray film (double-sided).
14. Developer (Kodak DX80).
15. Fixer (Kodak FX40).
16. Equipment: X-ray cassettes, intensification screens, −70°C freezer, dark room.

17. Radioactive ink can be produced by mixing a small amount of $^{32}P$- (or $^{35}S$-) labeled material with a waterproof black ink. The amount of isotope needs to be adjusted to conform to the activity of the material being autoradiographed.

### *For the Alternative Hybridization Procedure*

18. Denhardt's solution (50 ×):
    1% Bovine serum albumin (nuclease-free)
    1% Ficoll (molecular weight, 400,000)
    1% Polyvinyl pyrrolidine (molecular weight, 40,000)
    This should be made up in water, filter sterilized and stored at −20°C. Immediately before use, it is diluted with SSC buffer and water in the following ratio: 2 mL Denhardt's (50 ×), 6 mL SSC (20 ×), and 12 mL water. This diluted stock is referred to as 5 × Denhardt's.
19. SDS wash solution: 0.1% SDS in 3 × SSC buffer.

## Method

### *DNA Binding to the Filter*

1. Plate the transformation mixture on a suitable medium and incubate until small colonies (ca. 0:5 mm) are visible. This will usually require 10–12 h incubation. If the growth temperature is not an important consideration, incubation overnight at 30°C rather than 37°C will provide the small colonies that are needed.
2. Label a dry sterile nitrocellulose filter with a soft pencil. Lay it carefully on the surface of the plate, with the numbered side in contact with the colonies. Ensure that contact is made over the whole surface, and that the filter becomes completely moistened. Mark the filter and the underlying agar by stabbing with a needle in two or three asymmetric positions.
3. Transfer the filter, numbered side uppermost, to a fresh agar plate containing the appropriate antibiotic and incubate for several hours to allow the colonies to

grow. At the same time, incubate the master plate for a few hours before storing it at 4°C.

4. Soak a pad of filter paper in 0.5*M* sodium hydroxide. Transfer the membrane onto this pad, colony side uppermost, ensuring that no air bubbles are trapped underneath the membrane. This will lyse the colonies and denature the DNA. Do not allow any liquid to flow onto the surface of the membrane.
5. After approximately 10 min, transfer the membrane to a pad of filter paper that has been well soaked in 1*M* Tris-HCl, pH 7.4. Remove the membrane after 2 min and transfer it to a fresh pad of filter paper soaked in the same buffer. Leave it in contact with this filter paper pad for another 2 min.
6. Transfer the membrane to a filter paper pad soaked in NaCl–Tris-HCl buffer. Leave it for about 15 min, then remove, and air dry the membrane.
7. Bake the dried membrane in a vacuum oven at 80°C for 2 h.

The filter now contains bound denatured DNA imprinted in the pattern of the colonies on the original master plate.

## *Hybridization*

1. Place the baked filter into a plastic bag and wet it with 5 mL of 50% formamide in 2 × SSC. Take care that air is not trapped in the nitrocellulose matrix.
2. Add 10 μL of 1*M* sodium hydroxide to the $^{32}$P-labeled probe and boil it for 10 min. Cool it on ice and neutralize it with 10 μL of 1*M* HCl.
3. Add the probe to the plastic bag containing the filter. Remove as much air as possible and seal the bag. Incubate at 37°C for 16–48 h.
4. Recover the filter and wash successively with 50% formamide in 2 × SSC (twice) and 2 × SSC (twice). Each wash should be for about 30 min, and should be with about 200 mL of solution. Carry out the formamide washes in a fume cupboard.
5. Air dry the filter and mount on a sheet of filter paper (e.g., 3MM) for autoradiography.

## Autoradiography

1. Mount the filter(s) on a conveniently sized piece of Whatman 3MM paper. The position and orientation of the filters can be marked with radioactive ink.
2. Using the darkroom, place a sheet of X-ray film next to the sample in a light-proof cassette. Fix the film to the sample by a few pieces of tape at the extreme edges of the film. If an intensification screen is used, the film is sandwiched between the sample and the screen. The shiny side of the screen must face towards the X-ray film. Label the outside of the cassette. If the cassette is not absolutely light-proof, it should be wrapped in a heavy-duty black plastic bag.
3. Store at room temperature, or at −70°C if an intensification screen is used. The exposure time required for a particular experiment is dependent on a number of factors such as the specific activity of the probe and the number of copies of the particular DNA sequence, and is often best determined by trial and error; overnight exposure is often sufficient.
4. Take the cassette to the darkroom. Using the safelight, open the cassette and place the film in a tray of diluted developer (1:4 dilution) for about 4 min or until sufficiently developed.
5. Wash the film briefly in water, and then place it in a second tray containing fixer (1:4 dilution) for about 4 min. Wash the film extensively with running tap water and hang it up to dry.
6. Examine the film. A positive result will be shown as a dark spot corresponding to the position of a particular colony on the original master plate. That colony can then be picked for subculture, storage, and further testing.

# Notes

1. The above procedure is suitable for a situation where 100–200 colonies are present on an 8.5 cm plate. Larger plates can also be used. If the colony density is higher than this, it will not usually be possible to align the autoradiogram accurately with the master plate; how-

ever, an area of the plate will be identified as containing a positive signal. The colonies in that area can then be recovered as a mixture, diluted, replated at a suitable density, and the hybridization and autoradiography repeated.

If plating out the transformation mixture gives rise to a number of plates, each containing only a few colonies, these can be picked with sterile toothpicks onto a nitrocellulose filter on the surface of a suitable medium, and also to a master plate, using the same grid. The colonies on the filter are then allowed to grow briefly before being lysed as described above.

2. Remember to boil the probe immediately before use—even if it has already been boiled on a previous occasion.
3. **CAUTION!** Formamide is a potent teratogen. Handle all solutions containing formamide with care; wear gloves.
4. An alternative hybridization procedure is available using Denhardt's solution. This is as follows:

   (i) Place the baked filter into a plastic bag with 5 mL of 5 × Denhardt's solution. Seal the bag and incubate at 65°C for 3–4 h.
   (ii) Recover the filter, and place in another plastic bag with 5 mL of 5 × Denhardt's containing the boiled probe. Seal the bag and incubate at 65°C for 12–16 h.
   (iii) Recover the filter and wash with 200 mL of 0.1% (w/v) SDS in 3 × SSC at 65°C for 45–60 min. Repeat three times.
   (iv) Wash once with 3 × SSC at 65°C.
   (v) Air dry the filter and mount for autoradiography.

## References

1. Grunstein, M., and Hogness, D. (1975) Colony hybridization: a method for the isolation of cloned DNAs that contain a specific gene. *Proc. Natl. Acad. Sci.* **72,** 3961–3965.
2. Grunstein, M., and Wallis, J. (1979) Colony hybridization. In *Meth. Enzymology* **68,** 379–389. Academic Press, New York.

# Chapter 43

# Identification of Recombinant Phages by Plaque Hybridization

*J. W. Dale and P. J. Greenaway*

*Department of Microbiology, University of Surrey, Guildford, Surrey and Molecular Genetics Laboratory, PHLS Centre for Applied Microbiology and Research, Porton Down, Salisbury, Wilts., United Kingdom*

## Introduction

Phage vectors are often used rather than plasmids, particularly for the production of gene "banks" or "libraries." The plaques produced can be screened for the presence of specific DNA sequences by hybridization using a procedure similar to that used for colony hybridization (*see* Chapter 42). Plaque hybridization offers some advantages over colony hybridization, largely because the area of the filter to which the DNA is bound is smaller and more defined. A higher density of plaques can therefore be used, which in turn means that more plaques can be screened in a single hybridization—several thousand on an ordinary (8.5 cm) petri dish; using larger containers such as baking dishes or trays the number can run into hundreds of thousands in a single hybridization. In addition, since the location of the plaques is not disturbed by the process, nor are they smudged in the way that colonies can be, several replicates can be taken from the same plate. This means that the same set of plaques can be screened for hybridization to several different probes.

## Materials

1. Nitrocellulose membranes, of a size suitable for the plates to be screened. These membranes do not need to be sterile (unless the plaques are to be amplified), but they may need to be boiled to remove residual detergents and other contaminants that could contribute to high backgrounds after autoradiography.
   **DO NOT** heat dry nitrocellulose, except in a vacuum oven.
2. Whatman 3MM filter paper, about the same size as, or slightly larger than, the nitrocellulose membranes.
3. BBL agar:

| | | |
|---|---|---|
| Trypticase | 10 g | (Baltimore Biological Laboratories) |
| NaCl | 5 g | |
| Agar | 10 g | |

   Add water to 1 L; heat to dissolve the agar, then dispense in 100 mL amounts and sterilize by autoclaving. For the overlays, use 0.6% *agarose,* rather than agar.

4. 0.5*M* sodium hydroxide.
5. 1*M* Tris-HCl buffer, pH 7.4.
6. 1.5*M* NaCl in 0.5*M* Tris-HCl, pH 7.4.

7. Vacuum oven.
8. $^{32}P$-labeled DNA probe.

## Method

1. Plate the phage suspension (from transfection, in vitro packaging or from a phage stock) using BBL top layer agarose (rather than agar) on a BBL plate. Note that the plate should be dried thoroughly before pouring the overlay so that the top layer adheres well to the bottom layer.
2. Incubate the plates at 37°C until small plaques are visible (about 10 h).
3. Number a dry nitrocellulose membrane, using a soft pencil. Place the membrane carefully on the surface of the plate and mark both the membrane and the plate underneath at several asymmetrically located positions by stabbing it with a syringe needle dipped in ink. Leave the membrane in contact with the plate just long enough for it to become wet (30–60 s). Note that transfer of the phage is very rapid and the membrane must therefore not be moved after the initial contact has been made.
4. Remove the membrane with a pair of blunt forceps, taking care not to disturb or remove the top layer.
5. Place the membrane (plaque side up) onto a pad of filter paper soaked in 0.5*M* sodium hydroxide.
6. After about 10 min, transfer the membrane onto a pad of filter paper soaked in 1*M* Tris-HCl, pH 7.4. After 2 min, transfer the membrane to a second pad of filter paper soaked in the same buffer. Leave it for a further 2 min.
7. Remove the membrane and place it on a pad of filter paper that has been well soaked in 1.5*M* NaCl, 0.5*M* Tris-HCl, pH 7.4.
8. After approximately 15 min, remove the membrane and allow it to air dry.
9. Bake the dried filter in a vacuum oven at 80°C for 2 h.

10. Hybridize the filter with the appropriate $^{32}$P-labeled DNA probe as described in Chapter 42.

## Notes

1. It is important to avoid removing the overlay together with the filter. Drying the plate thoroughly before pouring the overlay encourages the top layer to stick to the plate rather than to the filter. The use of agarose rather than agar for the top layer also helps in this respect. If any of the top layer does stick to the filter it must be removed by careful washing.
2. Improved results can be obtained by amplification of the plaques before hybridization (2). This is achieved by immersing the filter in a diluted suspension of the indicator organism, and then allowing it to dry before transferring plaques to it. After transfer, the filter is placed on a fresh agar plate and incubated for 6–12 h. The filter will then carry a bacterial lawn with plaques and can be denatured and hybridized as described above. This procedure increases the amount of DNA bound to the filter, thereby increasing the strength of the signal arising from the positive clones as compared to the background. It also reduces the exposure time needed for autoradiography.

## References

1. Benton, W. D., and Davis, R. W. (1977) Screening lambda gt recombinant clones by hybridization to single plaques in situ. *Science* **196,** 180–182.
2. Woo, S. L. C. (1979) A sensitive and rapid method for recombinant phage screening. *Meth. Enzymology* **68,** 389–395.

# Chapter 44

# Plasmid Screening Using Single Colony Lysates

***J. W. Dale and P. J. Greenaway***

*Department of Microbiology, University of Surrey, Guildford, Surrey and Molecular Genetics Laboratory, PHLS Centre for Applied Microbiology and Research, Porton Down, Salisbury, Wilts., United Kingdom*

## Introduction

There are a number of methods available for screening either potential recombinant clones or natural isolates for the possession of plasmids. The method described in this chapter is based on that described by Sherratt (*1*) and is the simplest and most rapid of these. The name arises from the fact that the technique was devised to yield results using a single colony that can be taken directly from an appropriate agar plate. Although this procedure works best for high copy number plasmids, it can also be used to detect plasmids in environmental isolates, for example.

The technique involves resuspending a bacterial colony in buffer and then lysing the cells by heating in the presence of SDS. The crude lysate is loaded directly onto an agarose gel; after electrophoresis, the DNA bands are visualized in the normal way by staining with ethidium bromide. Both low and high molecular weight plasmids can be detected and sized, although some plasmids of intermediate size will be obscured by the heavy band of chromosomal DNA.

## Materials

1. Flat-ended toothpicks, sterilized by autoclaving.
2. Electrophoresis buffer: 0.04*M* Tris, 0.02*M* sodium acetate, 1 m*M* EDTA, adjusted to pH 8.2 with acetic acid. This buffer is normally made up at 10 × strength, autoclaved, and stored at 4°C. Dilute immediately prior to use.
3. Lysis mix: 5% sodium dodecyl sulfate, 0.05% bromophenol blue, 50% glycerol (or 10% Ficoll) in electrophoresis buffer.
4. Agarose, ethidium bromide.

## Method

1. Prepare a 1% agarose solution in electrophoresis buffer and pour a vertical gel.
2. Using the flat end of a toothpick, scrape a bacterial colony (enough to cover the end of the toothpick) off a plate and resuspend in 200 μL of electrophoresis buffer by rotating the toothpick rapidly between thumb and forefinger.
3. Add 50 μL of lysis mix and heat the tubes in a 65°C waterbath for 30 min. Vortex each tube for 20 s and return it to the waterbath until you are ready to load the samples.
4. Load 50 μL of each sample, while hot, into a slot in the gel.

5. Run the gel at 50 V for 5 h. Longer runs may be necessary for large plasmids.
6. Dismantle the gel, stain with ethidium bromide (0.5 μg/mL final concentration), and visualize the bands on a transilluminator. A heavy band of chromosomal DNA will be seen in the top third of the gel (in a position equivalent to a plasmid size of 10–15 Md) and a double band of RNA towards the bottom. Low molecular weight plasmids will run faster than the chromosomal DNA band; large plasmids will run behind this band.

## Notes

1. The crude lysates contain a high concentration of DNA and other cell components, which can make them very viscous, thus causing the loading of the gel to be very difficult. It is essential therefore to shear the mixture thoroughly by extensive vortex mixing, and to make sure that the lysates are hot when loading. If problems occur, repeat the vortex mixing stage.
2. The size of the plasmids can be determined by comparison with known plasmid standards treated in the same way. This, however, assumes that all the plasmids are present in a supercoiled configuration. A second band, moving more slowly, may also be seen, representing nicked (open circular) molecules.
3. Do not attempt to calibrate the gel using a lambda DNA restriction enzyme digest; this is only appropriate for linear DNA molecules.
4. The sensitivity of the method is variable. For multicopy plasmids, a single good-sized colony provides sufficient material. Large (single copy) plasmids are often more difficult to see since they may be obscured by the tailing of the chromosomal DNA. They are usually found as a very thin and comparatively faint band. For detection of low copy number plasmids, or if the colonies are small, it will be necessary to pick several colonies or even to streak the colony onto a fresh agar plate to provide sufficient material.

5. Alternative procedures are available that involve lysing the cells directly in the wells of the agarose gel using alkaline SDS (*see* ref. 2, for example). One advantage of doing this is that the chromosomal DNA is denatured and therefore does not run into the gel; it is thus prevented from obscuring plasmid bands. However, for most applications the procedure described above is simpler and more reliable.
6. There are a number of limitations to this procedure, the most serious of which is that the DNA cannot be cut with restriction enzymes. For most routine purposes, a small-scale plasmid preparation by the method of Birnboim and Doly (*3; see* Chapter 27) is preferable. The extra time involved is justified by the important additional information derived by analysis of the restriction enzyme digests.

## References

1. Sherratt, D. (1979) Plasmid vectors for genetic manipulation *in vitro*. In *Biochemistry of Genetic Engineering* (Garland, P. B. and Williamson, R., eds.), pp. 29–36. The Biochemical Society, London. (Biochemical Society Symposium No. 44).
2. Bidwell, J. L., Lewis, D. A., and Reeves, D. S. (1981) A rapid single colony lysate method for the selective visualisation of plasmids in Enterobacteriaceae, including *Serratia marcescens*. *J. Antimicrob. Chemother.* **8,** 481–485.
3. Birnboim, H. C., and Doly, J. (1979) A rapid alkaline extraction procedure for screening recombinant plasmid DNA. *Nucleic Acids Res.* **7,** 1513–1523.
4. Holmes, D. S., and Quigley, M. (1981) A rapid boiling method for the preparation of bacterial plasmids. *Anal. Biochem.* **114,** 193–197.
5. Barnes, W. M. (1977) Plasmid detection and sizing in single colony lysates. *Science* **195,** 393–394.

# Chapter 45

# Immunological Detection of Gene Expression in Recombinant Clones

*J. W. Dale and P. J. Greenaway*

*Department of Microbiology, University of Surrey, Guildford, Surrey and Molecular Genetics Laboratory, PHLS Centre for Applied Microbiology and Research, Porton Down, Salisbury, Wilts., United Kingdom*

## Introduction

After ligation and transformation, a number of clones will (it is hoped) be obtained that can be identified as recombinants by the occurrence of insertional inactivation, by analysis of small-scale plasmid preparations, or more specifically by *in situ* hybridization. An additional strategy that is superficially attractive, but not without considerable problems, is to look directly for expression of the cloned gene(s). This may be applicable in primary cloning if no specific DNA probe is available, but is more generally used in subsequent studies of expression of specific foreign genes.

In some cases, it may be possible to apply a rapid test for expression of a functional enzyme, e.g., by using a chromogenic substrate or by genetic complementation. This has several disadvantages:

(i) The method adopted will be specific to a particular enzyme, so that a new procedure must be devised for each gene being studied.
(ii) It is not always easy to devise ways of testing large numbers of colonies rapidly (it is important to remember that this may involve screening thousands or tens of thousands of clones).
(iii) Tests of this kind rely on significant levels of enzyme activity and are therefore not always sufficiently sensitive to detect the very low levels of gene product that may be formed in a recombinant clone.

The use of an immunological screening procedure, such as that described by Broome and Gilbert (*1*), on which this protocol is based, circumvents many of these disadvantages. The procedure is applicable to any gene product that is antigenic, and that can be obtained in a sufficiently pure form to produce good quality antiserum; large numbers of clones can be screened by this technique, which is analogous to *in situ* colony hybridization; and the method is sensitive enough to detect the required gene product at extremely low levels (only a few molecules per cell being needed).

The basis of the technique is as follows. A polyvinyl disc is coated with a specific antibody (IgG). The recombinant colonies are lysed using a virulent phage or chloroform, and the IgG-coated disc is placed in contact with the lysed colonies. Any molecules with the desired antigenic determinants are in this way transferred to the plastic disc in a replica of the position of the original colonies. These spots of antigen can then be located by using radioactively labeled IgG followed by autoradiography.

# Materials

## For Purification of IgG

1. Antiserum prepared by immunizing a rabbit with the purified antigen.
2. 25 m*M* sodium phosphate, pH 7.8, containing 1% glycerol.
3. DEAE-cellulose (Whatman DE52) equilibrated with the above buffer.
4. 25 m*M* sodium phosphate, pH 7.3, 0.1*M* NaCl, 1% glycerol.

## For Iodination of IgG

5. 0.5*M* potassium phosphate, pH 7.5.
6. Carrier-free $Na^{125}I$
7. 10 m*M* chloramine T (freshly prepared).
8. 10 m*M* potassium metabisulfite.
9. 10 m*M* potassium iodide.
10. PBS.
11. PBS containing 2% normal rabbit serum.
12. PBS containing 10% normal rabbit serum.
13. Fume cupboard.
14. Sephadex G-50 column.

## For Coating the Discs

15. Polyvinyl discs: PVC double-glazing plastic sheets obtainable from home repair centers; cut into discs approximately 8.5 cm diameter to fit into standard petri dishes. Other sizes and shapes can be used, e.g., for large square petri dishes, if more colonies are to be screened.
16. Nylon netting discs (about the same size as the PVC discs) can be cut from net curtain material.
17. Wash buffer: phosphate buffered saline (PBS) containing 2% (w/v) normal rabbit serum, 0.1% (w/v) bovine serum albumin.

18. 0.2*M* sodium bicarbonate (pH 9.2).
19. Glass petri dishes.

### For the Detection Procedure

20. L agar plates.
21. Plate lysate of lambda vir.
22. BBL agar plate.
    (*see* Chapters 29 and 30 for details)
23. 85 mm millipore filters.
24. Wash buffer: phosphate buffered saline (PBS) containing 2% (w/v) normal rabbit serum, and 0.1% (w/v) bovine serum albumin.
25. X-ray film, cassettes, intensification screens.

## Method

### Purification of IgG

1. Precipitate the IgG fraction from the immune antiserum with ammonium sulfate (50% saturation). Redissolve the precipitate in phosphate buffer (pH 7.8) containing 1% glycerol and apply to a DEAE-cellulose column equilibrated with the same buffer.
2. Pool the fractions containing the bulk of the flow-through material and precipitate the protein by adding ammonium sulfate to 40% of saturation.
3. Resuspend the pellet in one-third of the original serum volume in the phosphate–NaCl–glycerol buffer and dialyze against the same buffer. Remove any residual precipitate by centrifugation and store in aliquots at −70°C.

### Iodination of IgG

1. To 50 μL of purified IgG (approximately 5 mg/mL) add 2 mCi of carrier-free $Na^{125}I$, and 20 μL of chloramine T.
2. Mix and incubate at room temperature for 3 min.

3. Stop the reaction by adding 40 μL of potassium metabisulfite and 120 μL of potassium iodide.
4. Add 200 μL of PBS containing 2% normal rabbit serum and apply to a column of Sephadex G-50 (15 × 0.7 cm diameter) equilibrated with this buffer.
5. Elute with PBS containing 2% normal rabbit serum and monitor the column effluent. The first radioactive peak contains iodinated IgG, the second is unincorporated label.
6. Dilute the labeled IgG fraction to 5 mL with PBS containing 10% normal rabbit serum, sterilize by filtration, divide into aliquots and store at −70°C.

*Note*: Over-iodination of the IgG can result in the loss of antigen binding capacity.

## Coating Polyvinyl Discs

1. Wash the polyvinyl discs with ethanol, then with distilled water and allow them to dry. Make sure the discs are flat; flatten them if necessary by placing them between sheets of filter paper and putting a suitable weight on top.
2. Place 10 mL of 0.2*M* $NaHCO_3$ (pH 9.2) containing approximately 60 μg IgG/mL (unlabeled) in a glass petri dish. Mark one side of the disc and float it on the surface of the IgG solution (marked side downward).
3. Remove the disc after 2 min and wash both sides by gently swirling in a dish with 10 mL of cold wash buffer. Remove the disc and repeat the wash.
4. Give the disc a final rinse in distilled water and allow to dry. There should be no visible deposit on the discs after drying.

*Note*: The original IgG solution should be good for about 10 discs. Use the discs immediately after coating.

## RIA Procedure

1. Place a millipore filter on the surface of an L-agar plate. Inoculate the filter by replica plating or by pick-

ing colonies with sterile toothpicks. Incubate the plate with the filter on it at 37°C overnight.

2. Prepare an L-agar plate lysate of lambda vir. Transfer the millipore filter (face down) to this plate, allow the phage to absorb for 2 min, then transfer the filter (face up) to a BBL plate for incubation at 37°C for 4–6 h. An alternative procedure is to place a filter paper disc in a petri dish lid and add a few drops of chloroform. The L-agar plate with the millipore filter is then inverted over the chloroform pad and left for about an hour to lyse the colonies. (To make extra sure, use both lysis procedures!).
3. Gently place an IgG-coated PVC disc (marked side downwards) in contact with the lysed colonies. Smooth out any air bubbles and leave the plate at 4°C for at least 3 h (preferably overnight) for immunoadsorption of antigen onto the antibody-coated solid phase. If this is done on the BBL plate the millipore filter remains moist.
4. Label the plastic disc and the millipore filter unambiguously so that the colonies can be easily lined up. Avoid the use, wherever possible, of pentel or magic marker since some inks of this nature may expose X-ray film.
5. Separate the plastic disc from the millipore membrane. Wash the disc thoroughly with ice-cold wash buffer to remove nonspecifically adhered cellular material. The remains of the bacterial colonies may be very sticky and not easily washed off. As a last resort, *gently* squirt the wash buffer over the surface of the disc using a Pasteur pipet.
6. Dilute the labeled IgG in wash buffer to approximately $5 \times 10^6$ cpm/mL. Place a nylon disc in the bottom of a glass petri dish and add 1.5 mL of the diluted labeled IgG. Place the polyvinyl disc, coated side down, onto the nylon disc. Cover the PVC disc with another piece of nylon net and repeat the layering process for as many times as convenient. Make sure that each plastic disc has sufficient isotope in contact with its coated side.

7. Incubate for 16–24 h at 4°C and then wash each disc twice with cold wash buffer, using gentle swirling. Then wash twice more with water.
8. Lightly blot or shake the discs to remove water droplets and allow to dry at room temperature.
9. Fasten the discs (coated side up) to a sheet of filter paper in a light-proof cassette, using small pieces of sticky tape.
   In the darkroom, using a safelight, place a sheet of X-ray film in contact with the discs and put an intensification screen (DuPont Cronex Lightning Plus) on top of the film. Close the cassette (if it is not absolutely light-proof, wrap it thoroughly in a heavy duty black plastic bag) and store it at −70°C.
10. Develop and fix the autoradiogram as usual. The presence of positive clones will be indicated by black spots on the film.

# Notes

1. One of the advantages of this procedure is that it does not require the presence of an intact functional gene product. Under some circumstances, partial or fused products may also be antigenic and will then be detected. Some expression vectors are specifically designed to give rise to fused gene products.
2. Additional details and alternative procedures for the purification and iodination of IgG will be found in standard immunological manuals (*see* refs. *2* and *3*, for example). Alternative versions of the detection procedure are also available (*see* refs. *4* and *5*).
3. Many rabbits (and other laboratory animals) possess nonspecific coliform antibodies. Suitable controls must therefore be incorporated. If necessary, the antiserum can be absorbed out with an extract from the plasmid-free transformation strain.

## References

1. Broome, S., and Gilbert, W. (1978) Immunological screening method to detect specific translation products. *Proc. Natl. Acad. Sci. USA.* **75,** 2746–2749.
2. Hudson, L., and Hay, F. C. (1980) *Practical Immunology.* Blackwell Scientific Publications, Oxford, UK.
3. Bolton, A. E. (1977) *Radioiodination Techniques.* Amersham International Ltd.
4. Clarke, L., Hitzeman, R., and Carbon, J. (1979) Selection of specific clones from colony banks by screening with radioactive antibody. *Meth. Enzymology* **68,** 436–442.
5. Erlich, H. A., Cohen, S. N., and McDevitt, H. O. (1979) Immunological detection and characterization of products translated from cloned DNA fragments. *Meth. Enzymology* **68,** 443–453.

# Chapter 46

# The Isolation of Minicells

**_Wim Gaastra and Per Klemm_**

*Department of Microbiology, The Technical University of Denmark, Lyngby, Denmark*

## Introduction

The term minicell was introduced by Adler et al. (*1*) for the small spherical cells produced by abnormal cell division at polar ends of certain *Escherichia coli* cells. The phenomenon itself however, had already been described as early as 1930 (2) for a strain of *Vibrio cholera*. These minicells do not contain any chromosomal DNA, but they do inherit in most cases the plasmid DNA present in the parental strain. Since plasmid-containing minicells are able to synthesize DNA, RNA, protein, and other cell components, they are specifically suited for the in vivo analysis of the gene products encoded for by their plasmid contents, in the absence of gene products encoded for by the host genome. In this chapter the isolation of minicells will be described, whereas the identification of gene products in minicells will be described in Chapter 48. Other methods currently used to analyze the gene products of a particular stretch of DNA are the maxicell system (Chapter 47) and the various DNA-dependent in vitro protein synthesis systems (Chapter 19).

Obviously the DNA to be analyzed has to be introduced into the minicell producing strain before isolation of the minicells.

## Materials

1. A minicell-producing strain, for example, the *E.coli* K12 strain, P 678–54 ($F^-$, *thr, ara, leu, azi, ton*A, *lac*Y, $T6^s$, *min*A, *min*B, *gal, rps*L, *mal*A, *xyl, mtl, thi, sup*) (*see* ref. *3*) or DS 410 (prototroph, λS) (*see* ref. *1*).
2. Buffered Saline Gelatin (BSG) (ten times concentrated): 0.85% NaCl, 0.03% $KH_2PO_4$, 0.06% $Na_2HPO_4$, and 100 μg/mL gelatin
   The pH of this solution should be 7.7. The gelatin is dissolved by warming the solution to 100°C in a boiling water bath. The 10 × solution is stored in the cold.
3. Growth medium for the minicell-producing strain. A 300 mL volume of growth medium per strain is used, to which the appropriate antibiotic is added when the plasmid to be analyzed encodes for antibiotic resistance. Usually Brain Heart Infusion (BHI) or L-Broth supplemented with methionine is used (2 mL of 2% methionine to 300 mL of L-broth).
4. Sucrose gradients (two or three per strain).
   The sucrose gradients are made by making a 20% sucrose solution in BSG, filling sterile polycarbonate centrifugation tubes (80 and 40 mL, *see* Method) and freezing them at −20°C for 3–4 h. The gradient is formed by allowing the tubes to thaw overnight at 4°C.
5. Disposable 10 mL syringes plus needles (two or three per strain).
6. Polycarbonate centrifugation tubes: 80 mL, one per strain; 40 mL, two per strain.

## Method

1. Inoculate 300 mL of growth medium with the minicell-producing strain containing the plasmid to be analyzed. Add appropriate antibiotics if necessary. Grow overnight at 37°C in a shaking incubator.

2. Harvest the cells by centrifugation (10 min, 12,000 rpm at 4°C).
3. Resuspend the cells in 10 mL of BSG (1 ×). Resuspending should be done vigorously. Either use a 5 mL plastic syringe or a 10 mL pipet to disrupt the pellet while vortexing. Suck the suspension in and out of the pipet 4–5 times.
4. Harvest the cells by centrifugation (10 min, 12,000 rpm at 4°C).
5. Resuspend the cells in 4 mL of BSG (1 ×). Use a Vortex and a 5 mL pipet. Vortexing should be continued for at least 2 min.
6. Layer the cell suspension on top of a 80 mL sucrose gradient. Spin for 30 min at 3700 rpm in a MSE tabletop low-temperature centrifuge (Chillspin) at 4°C. Do not use the brake.
7. Remove the upper band, containing the minicells, from the gradient with a 10 mL syringe. Unjam the syringe plunger before taking the band out, otherwise disruption of the gradient and mixing with whole cells may occur. For this reason the last 1/4–1/5 of the minicell band is also not taken from the gradient.
8. Dilute the minicells with 1 × BSG (4°C) and spin for 10 min at 20,000 rpm (4°C).
9. Resuspend the pellet in 1 mL BSG, using a Vortex and a 1 mL pipet.
10. Layer the minicell suspension on top of a 40 mL sucrose gradient. Spin the gradient for 30 min in an MSE Chillspin at 3700 rpm at 4°C. Do not use the brake.
11. Repeat steps 7, 8, and 9.
12. After the third sucrose gradient, the minicell pellet is resuspended in label medium (*see* Chapter 48) until it reaches an optical density at 660 nm of 0.4.

# Notes

1. The minicell suspension can be checked for purity by plating out 20 μL of the suspension on an agar plate containing rich medium. These 20 μL should contain less than 20 colony forming cells. They contain however, approx. 20 × $10^8$ minicells.

2. The minicells can be used directly for the labeling of proteins or stored at −70°C for later use.
3. To store minicells, add 500 μL of sterile glycerol to 1 mL of minicell suspension and store at −70°C. Before using stored cells carry out the following procedure:

   (a) Spin the suspension for 20 min at 12,000 rpm and 4°C. Discard 1 mL of the supernatant (500 μL is thus left).
   (b) Add 1.5 ml of label medium, resuspend, spin, and discard 1.5 mL of supernatant. Repeat 4 times.
   (c) Discard the supernatant completely, add 2 mL label medium, and wash once with 2 mL of the same medium.
   (d) Finally resuspend in 1 mL of label medium.

4. It is clear that the method described, especially the centrifugation of the sucrose gradients, has to be adapted to the centrifuges and the rotors present in the laboratory. Spinning the gradients in a SW 25.7 rotor at 5000 rpm for 20 min for instance, gives a pellet of the minicell-producing cells and a band containing minicells in the middle of the tube.
5. Instead of preparing the sucrose gradient by freezing and thawing, step gradients can also be used. These gradients are made by layering 10 mL of a 5% sucrose solution in BSG on top of 30 mL of a 20% sucrose solution in BSG. On these gradients 2 mL of cell suspension is layered instead of 4 mL.

## References

1. Adler, H. I., Fisher, W. D., and Hardigree, A. A. (1967) Miniature *Escherichia coli* cells deficient in DNA. *Proc. Natl. Acad. Sci. USA* **57,** 321–326.
2. Gardner, A. D. (1930) Cell division, colony formation and spore formation. In *A System of Bacteriology in Relation to Medicine* (Fildes, P., and Ledingham, J. C. G., eds.) HMSO London, **1,** 159–176.
3. Dougan, G., and Sherratt, D. (1977) The transposon Tn 1 as a probe for studying Col El structure and function. *Mol. Gen. Genet.* **151,** 151–166.

# Chapter 47

# The Isolation of Maxicells

***Wim Gaastra and Per Klemm***

*Department of Microbiology, The Technical University of Denmark, Lyngby, Denmark*

## Introduction

The maxicell system (*1*) is based on the following observation: The number of lesions induced with UV light in a DNA molecule is proportional with the size of the molecule and the dose of the UV light. Since most plasmids are less than 1/1000 the size of the chromosome and are usually present in 10–100 copies per cell, it is possible to irradiate a culture of a plasmid-containing strain in such a way that every single chromosome is damaged at least once, whereas at the same time a substantial number of plasmid molecules remain undamaged.

Since in wild-type *E. coli* cells there are enzymes capable of repairing these UV damages, it is necessary to use a strain with mutations that inactivate these repair systems.

In such a strain, every damaged DNA molecule will be completely degraded. However, when these cells contain multicopy plasmids that have not been hit by the UV irradiation, the number of plasmids increases by a factor of 10 in about 6 h. In other words, plasmids are still being replicated in these cells and their gene products can be analyzed in the absence of chromosomally encoded gene products.

## Materials

1. *E. coli* strain CSR 603: *thr*-1, *leu*B-6, *pro*A-2, *arg*E-3, *phr*-1, *rec*A-1, *uvr*A-6, *thi*-1, *ara*-14, *lac*Y-1, *gal*K-2, *xyl*5, *mtl*-1, *rps*L-31, ($str^R$), *tsx*-33, *sup*E-33
2. AB medium or M9 medium. To prepare AB medium, take A medium (100 mL) and B medium (10 mL) and make to 1 L with distilled water.

| | | |
|---|---|---|
| *A medium* (pH 6.4) (10×): | $(NH_4)_2SO_4$ | 20 g/L |
| | $Na_2HPO_4 \cdot 2H_2O$ | 60 g/L |
| | NaCl | 30 g/L |
| | $KH_2PO_4$ | 30 g/L |
| *B medium* (100×): | $MgCl_2 \cdot 6H_2O$ | 203 g/L |
| | $CaCl_2 \cdot 2H_2O$ | 14.7 g/L |
| | $FeCl_3 \cdot 6H_2O$ | 2.70 g/L |
| *M9 medium* (pH 7.4): | $Na_2HPO_4$ | 6 g/L |
| | $KH_2PO_4$ | 3 g/L |
| | NaCl | 0.5 g/L |
| | $NH_4Cl$ | 1 g/L |
| *Sterilize separately before adding* | 1*M* $MgSO_4$ | 2 mL/L |
| | 20% glucose | 10 mL/L |
| | 1*M* $CaCl_2$ | 0.1 mL/L |

3. UV source, for example a GE 15 W germicidal lamp or equivalent.

## Method

1. A 20 mL culture of strain CSR 603 that has been transformed with the plasmid of which the gene products have to be analyzed, is inoculated from a fresh overnight culture in a 200 mL flask and grown at 37°C in AB medium (or M9 medium) containing 0.2% glucose, 1% casamino acids and 0.1 µg/mL thiamine.
2. Start the culture with an optical density at 450 nm of 0.04. Follow the OD every 60 min.
3. At $OD_{450}$ = 0.6–0.8, 10 mL of the culture is poured into a sterile petri dish and irradiated with UV light

for 10 s from a distance of 60 cm with a GE 15 W germicidal lamp.

4. The irradiated culture is transferred to a 35 mL sterile centrifuge tube and incubated while shaking for 1 h at 37°C.
5. Add 200 μg/mL cycloserine and continue the incubation over night.
6. Harvest the culture and wash twice with AB medium.
7. Resuspend in 3 mL of AB medium.
8. Transfer 1.5 mL to a big Eppendorf tube and pellet the cells.
9. Wash the cells with 1.5 mL of AB medium.
10. Resuspend in 1 mL of AB medium supplemented with 0.2% glucose, 50 μg/mL of the required amino acids, and 0.1 μg/mL thiamine.
11. Incubate while shaking at 37°C for 1 h.
12. Add the appropriate radioactive precursor (*see* Chapter 48).

# Notes

1. The success of the maxicell system is critically dependent on the different target size of the plasmids and the chromosome. The maximum size of plasmids that can be studied in maxicells is not exactly known, but plasmids with a molecular weight of up to $1 \times 10^7$ can be used.
2. The method of UV irradiation needs some attention. It is easy to irradiate the cells too much or too little. This is of course dependent on the UV lamp one chooses, as well as the distance of the cells from the UV source. It is best to determine this empirically.
3. The petri dish should be shaken while irradiating the culture.
4. Cycloserine is an antibiotic that, like penicillin, will lyse all cells that grow, i.e., it ensures the removal of any surviving cells.
5. It is possible to store maxicells at −70°C in glycerol, in the same way as described for minicells (*see* Chapter 46).

## References

1. Sancar, A., Hack, A. M., and Rupp, W. O. (1979) A simple method for identification of plasmid coded protein. *J. Bacteriol.* **137**, 692–693.

# Chapter 48

# The Identification of Gene Products in Minicells and Maxicells

***Wim Gaastra and Per Klemm***

*Department of Microbiology, The Technical University of Denmark, Lyngby, Denmark*

## Introduction

Having introduced a recombinant DNA molecule to be analyzed into either a minicell or a maxicell, it is then possible to detect the gene products encoded for by this DNA by incubation of a minicell or maxicell preparation with radioactively labeled precursors of RNA and protein; uridine or uracil in the case of RNA and amino acid mixtures, or specific amino acids in the case of proteins. Since neither minicells nor maxicells contain chromosomal DNA, if one allows a preincubation period long enough to get rid of any long-lived messengers transcribed from the chromosomal DNA, the radioactivity will only be incorporated into gene products encoded for by the recombinant DNA present in the cells. The method described here is based on a modified procedure reported by Mooi et al. (*1*).

## Materials

1. A pure minicell preparation (*see* Chapter 46)
2. Label medium for minicells: 20 mL M9 medium (10 × concentrated); 2 mL 0.1*M* $MgSO_4$; 2 mL 0.01*M* $CaCl_2$; 0.2 mL vitamin $B_1$ (4 mg/mL); 1.6 mL 50% glucose; and 173 mL $H_2O$.
3. Label medium for maxicells: AB medium (*see* Chapter 47) *or* M9 medium each containing 0.2% glucose, 50 μg/mL of the required amino acids and 0.1 μg/mL thiamine.
4. M9 medium (10 × concentrated): $Na_2HPO_4$, 60 g/L; $KH_2PO_4$, 30 g/L; NaCl, 5 g/L; and $NH_4Cl$, 10 g/L.
5. SDS buffer: 50 m*M* Tris-HCl, pH 6.8; 1% SDS; 1% B-mercaptoethanol; and 5% glycerol.
6. 10% TCA solution.
7. Ice-cold acetone.

## Method

1. Take 1 mL of minicell or maxicell suspension in the appropriate label medium (OD 660 nm = 0.5–0.6) and incubate for 15 min at 37°C.
2. Add radioactively labeled precursors, e.g., 50 μCi of $^{35}S$ methionine or $^{14}C$- or $^{3}H$-labeled amino acid(s). Radioactive uridine or uracil should be used when RNA is the product to be analyzed.
3. Incubate for 1–2 h at 37°C. Occasionally overnight incubation may improve the incorporation of label.
4. Stop the labeling by spinning down the cells (2 min in an Eppendorf centrifuge).
5. Remove the supernatant (very radioactive) and resuspend the pellet in 160 μL of SDS buffer.
6. Incubate for 3 min in a boiling water bath.
7. Add 1 mL of ice-cold acetone and incubate on ice for 30 min.
8. Centrifuge in an Eppendorf centrifuge for 3 min.
9. Remove the supernatant (radioactive) and dry the pellet in a vacuum desiccator.
10. Resuspend the pellet in 50 μL of sample buffer. This can either be the buffer described for SDS gel electro-

phoresis in Chapter 6, Vol. 1, or the buffer for RNA electrophoresis described in Chapters 11 or 12.

11. Take out 2 μL from the suspension and precipitate it with 10% TCA to determine the amount of radioactivity that has been incorporated. The TCA precipitation is carried out as follows:
    (a) Add 2 μL of sample to 5 mL of 10% TCA. Incubate on ice for 1 h.
    (b) Filter the sample through a suitable filter paper (e.g., GF/C). Wash twice with 5% ice-cold TCA and once with 20 mL of acetone.
    (c) Dry the filter at 120°C for 15 min.
    (d) Determine the amount of incorporation in a scintillation counter.
12. Analyze the products formed by gel electrophoresis and autoradiography/fluorography (*see* Chapter 6, Vol. 1, for proteins and Chapters 11 or 12 for RNA).

# Notes

1. It is of course possible to use more than 1 mL of minicell or maxicell suspension for labeling. In this case incubation in label medium is preferentially performed in a small glass vial in a shaking incubator. This usually gives better incorporation of the radioactive precursors.
2. Labeling can also be stopped by the addition of a 1000-fold excess of unlabeled amino acid(s).
3. When analyzing recombinant DNA molecules in minicells with respect to the gene products they encode for, it should be borne in mind that minicells differ in many respects from normal cells. Chromosomally encoded factors active in repression or induction of plasmid functions may be absent and lead to a certain variability in the expression of plasmid genes. The so-called signal peptidase that cleaves the leader sequence from a protein precursor, also seems to work less efficiently in minicells.
4. Not all genes are expressed in maxicells. Genes normally expressed at very low levels will also be expressed at low levels in maxicells. The reason for this can be either the presence of a specific repressor, an

ineffective promotor, or a poorly translated messenger RNA.

# References

1. Mooi, F. R., Harms, N., Bakker, D., and de Graaf, F. K. (1981) Organization and expression of genes involved in the production of the K88ab antigen. *Infect. Immun.* **32,** 1155–1163.

# Chapter 49

# Molecular Cloning in Bacteriophage Lambda and in Cosmids

***Claus Christiansen***

*Gensplejsningsgruppen, The Technical University of Denmark, Lyngby, Denmark*

## Introduction

Bacteriophage lambda contains a double-stranded DNA molecule of about 50 kilobases (kb). This molecule is linear in the phage particle and it possesses a set of single-stranded complementary ends (the *cos* region) by which the molecule is joined to form a circular molecule in the first step of the infection.

On the large left part of the genome are situated a number of genes producing essential components of the phage particle, while the right part codes for DNA replication and lysis of the host. The large part of the central genome (from the end of J to the region at $C_I$) of lambda is not essential for lytic infection and can be replaced (*1–3*). The DNA being inserted may originate from the *E. coli* host, as seen in classical in vivo lambda recombinants, or it may originate from foreign DNA inserted by enzymatic manipulations. Wild-type lambda or simple derivatives of lambda are not suitable for in vitro recombinant techniques for several reasons. For one, most restriction enzymes cleave lambda DNA into many fragments with destruction of essential regions as a consequence.

Insertion of a foreign fragment of DNA into lambda is not easily recognized without a selective marker. Lambda cloning vectors are therefore purposefully derived from wild-type lambda by selection for mutants with fewer restriction enzyme cleavage sites and by insertion of genetic markers that are eliminated or destroyed by insertion of foreign fragments in suitable sites.

The vectors are used in two essentially different ways. One is by insertion of the foreign DNA into a restriction enzyme site with subsequent inactivation of a gene (insertional cloning). The other is by replacement of a restriction enzyme fragment (a stuffer fragment) of the vector with the DNA to be cloned (replacement cloning). In this last case, gene inactivation is by deletion of the stuffer fragment. Insertional cloning will allow cloning from 0 to about 14 kb, whereas replacement will allow cloning of up to about 24.5 kb.

Of the several genes used for identification or selection of recombinants a few will be mentioned. Presence of *lac* or *ara* on the phage genome will give blue or red plaques on Xgal or MacConkey medium with arabinose, respectively. Deletion or inactivation of these genes in recombinants will allow the identification of the colorless plaques of recombinants. The genes *red* and *gam* are, in lambda wild-type phage, situated at about 33–34 kb of the phage genome in the dispensable region; ordinary bacteriophage lambda will not grow on P2 lysogens (**S**ensitive to **P2 I**nterference, the Spi$^+$ phenotype). Removal of *red* and *gam* will change the phage to Spi$^-$, i.e., recombinant phage will grow on a P2 lysogen, whereas nonrecombinants with functioning *red* and *gam* will not grow. This principle is used in the phage constructions lambda 1059 and lambda 47. Lambda 1059 is further modified to the EMBL line of vector phages (*3,5*).

Cosmids (*6*) are plasmids of the type usually used as vectors (fx. pBR322) with the lambda *cos* region inserted.

After proper cleavage this plasmid type may be ligated to large DNA fragments to form molecules of plasmid alternating with the large DNA fragments. These very large molecules can be packed into the bacteriophage lambda particle by in vitro systems provided the distance between *cos* regions is about 50 kb and the *cos* regions are oriented correctly. Upon infection of a bacterium, the molecule is circularized and survives as a normal recombinant plasmid with a very large insert. Antibiotic markers are used to identify recombinants.

Bacteriophage lambda and cosmid vectors are most frequently used to form recombinant DNA libraries to represent genomes. Specific recombinants can be searched for in these libraries by plaque hybridization (Chapter 43), by colony hybridization (Chapter 42), and by immunological methods (*see* Chapter 45). The experiments involved in a library formation are

(a) Choice and preparation of lambda or cosmid vector (*see* Note 1).
(b) Preparation of passenger DNA suitable for library formation by random fragmentation by partial enzyme digestion or shearing. In the last case linkers are fitted to the sheared DNA molecules after they are protected with the relevant restriction methylase. Cleavage of the linkers with the restriction enzyme corresponding to the methylase (most frequently EcoRI) completes the preparation of the passenger. Suitable passenger sizes are for lambda about 20 kb and for cosmids about 40 kb (*see* ref. 7, p. 280 ff.)
(c) Ligation of vector DNA to passenger DNA.
(d) Packaging of ligated DNA.
(e) Infection of cells.
(f) Library formation.

# Materials

1. *Lambda vector DNA*: Lambda 47.1 is chosen. It has 2 BamHI sites at 23580 and 30190 base pairs (bp) and a total length of 40620 bp. Removal of the stuffer fragment removes the genes *red* and *gam* (2,7).

2. *Bacterial hosts: E. coli* Q358 ($hsdR_k^-$,$hsdM_k^+$, supF,-ø80$^r$); *E. coli* Q359 ($hsdR_k^-$,$hsdM_k^+$, supF,ø80$^r$,P2); *E. coli* HB101 ($hsdR_b^-$ $hsdM_b^-$,$recA^-$,$ara^-$,pr $oA^-$,-lacY1,$Sm^+$, supE).
3. *Cosmid DNA*: pHC79 is among the most frequently used cosmids. It is constructed from pBR322 by inserting about 2 kb of lambda DNA from the region carrying *cos*. It is inserted in the region between the *tet* gene and the replication region of pBR322 (*6*).
4. *Passenger DNA*: The passenger DNA must be of high molecular weight. Before use it is cleaved partially with SauA3 or MboI and size fractionated to 20 kb (for libraries in lambda) or to 40 kb (for libraries in cosmids).
5. *High Salt Buffer*: 100 m*M* NaCl; 10 m*M* $MgCl_2$; 1 m*M* DTT, 50 m*M* Tris/HCl, pH 7.5
6. *TE Buffer*: 10 m*M* Tris/HCl, pH 8.0, 1 m*M* EDTA (sodium salt).
7. *Centrifugation buffer* (for sucrose gradients): 1.0*M* NaCl; 0.005*M* EDTA; 0.02*M* Tris/HCl, pH 8.0.
8. *SM Buffer*: 0.1*M* NaCl; 0.01*M* $MgCl_2$; 0.05*M* Tris/HCl, pH 7.5; gelatin, 2% (w/v).

# Methods

## *Method A: Cloning in Lambda*

1. Digest 10 μg of DNA from lambda 47.1 with 20 U of BamHI and 20 U of SalI restriction endonucleases for 2 h at 37°C. The total volume of the reaction should not exceed 100 μL. The two enzymes both work in high salt buffer.
2. Run a sample of the DNA on a 0.7% agarose gel to check for complete digestion.
3. Extract the DNA twice with an equal volume of phenol.
4. Prepare two 10–30% sucrose gradients in centrifugation buffer in SW60 tubes or equivalent (*see* Note 2).
5. Gently layer the DNA on top of the gradient, place in centrifuge and run for 55,000 rpm for 3 h at 10°C (the other tube is conveniently used for fractionation of passenger DNA).

6. Fractionate the gradients into 0.25 mL fractions and run 25 μL samples of such fractions on a gel. Use a sample of cleaved vector DNA as a reference. Identify the relevant fractions (those with the 10 kb left arm and those with the 23 kb right arm). Avoid those with fragments of the stuffer (4.8, 1.2, and 0.5 kb). Pool the fractions and precipitate with EtOH. Dissolve in 10 μL of TE buffer. Run 1 μL on a gel to check purity and concentration.
7. Mix 1 μg of arms with 0.5–0.7 μg of passenger DNA (molar ratio 1:1), then add ligation buffer and DNA ligase and ligate in a final volume of 10 μL (*see* Chapter 32). The ligation is done at 37°C for 1 h (to ligate the *cos* ends) and then at 14°C overnight (to ligate the BamHI ends of the vector to the SauA3 ends of the passenger) (*see also* Note 3).
8. Pack DNA (0.2 μg) into phage particles by an in vitro packaging reaction (Chapter 36) (*see also* Note 4). Dilute the packed phage to 0.2 mL with SM buffer.
9. Plate a 1/10 dilution series of the phage on Q359. A titer of $10^4$–$10^5$ recombinants/mL should be obtained. Extracts of the plates with confluent lysis can be used as recombinant library stocks. Prints can be taken from the more diluted plates on nitrocellulose filters and be used for screening by immunological reactions or by hybridization.

## Method B: Cosmid Cloning

1. Digest 5 μg of pHC79 with BamHI in high salt buffer.
2. Extract the DNA twice with phenol and precipitate with EtOH.
3. Mix 1 μg of cosmid DNA with 1 μg of passenger DNA (a molar proportion of 5:1) then ligate in a volume of 10 μL at 14°C overnight (*see* Chapter 32).
4. Pack the DNA into phage particles (*see* Chapter 36).
5. Infect HB101 with 100 μL of the phage solution, dilute with broth and incubate at 37°C for 90 min.
6. Plate a dilution series on ampicillin plates (50 μg $mL^{-1}$) and incubate overnight.

7. Check recombinants for tetracycline sensitivity and analyse as for other plasmid transformants (*see* Chapter 43).
8. The cosmid collection may be passed over strain BHB3064 (a lambda lysogen harboring a defective lambda phage. At 39°C the bacterium will pack cosmid copies without production of phage).

# Notes

1. Vectors are chosen according to desired principles for selection of recombinants, restriction enzyme and cloning capacity (*see* ref. *7*). A new set of vectors, the EMBL line (*5*) offers a polylinker sequence with different combinations of SalI, BamHI, and EcoRI recognition sequences at the ends of the spacer fragment. The stuffer fragment can be completely eliminated from participation in the ligation reaction by a double cleavage and a precipitation with isopropanol thus avoiding the sucrose gradient step in the methods. The last of the three restriction sites remain on the phage arms and can be used for exicion of the inserted DNA in analytical experiments.
2. This example is based on use of the SW60 rotor. Any swinging bucket rotor may be used provided the conditions are changed for it. Separation of fragments by gel electrophoresis and extraction of them from the gel may be preferred (*see* Chapter 10).
3. In ligation reactions, with the two lambda arms present, the stuffer fragment present, and a collection of passenger DNA fragments present, many different types of molecules can be formed. In library formation, high efficiency is desirable and under these circumstances removal or destruction of the stuffer fragment is advisable. When the collection of passenger molecules involves only a few types, the BamHI cleaved vector may be used directly. The stuffer fragment itself will not interfere because it is selected against by plating on the P2 lysogen, but stuffer fragments ligated to each other or to passenger fragments may be observed in the recombinant library.

4. It is essential that the packaging system is checked for efficiency with lambda DNA with ligated cohesive ends. Packaging systems may work wel with unligated DNA and not at all with DNA with ligated *cos* sites.

## References


1. Zehnbauer, B. A., and Blattner, F. R. (1982), Construction and Screening of Recombinant DNA Libraries with Charon Vector Phages, in *Genetic Engineering*, Vol. 4, Setlow, J. K., and Hollaender, A. eds., Plenum Press, New York pp. 249–279.
2. Brammer, W. J. (1982), Vectors Based on Bacteriophage Lambda, in *Genetic Engineering*, Vol. 3, Williamson, R., ed., Academic Press, London, pp. 53–81.
3. Hendrix, R. W., Roberts, J. W., Stahl, F. W., and Weisberg, R. A. eds. (1983), *Lambda II*, Cold Spring Harbor Laboratory, New York.
4. Karn, J., Brenner, S., and Barnett, L. (1983), New Bacteriophage Lambda Vectors with Positive Selection for Cloned Inserts, *Meth. Enzymol.* **101,** 3–19.
5. Frischauf, A.-M., Lehrach, H., Poustka, A. M., and Murray, N. (1983), Lambda Replacement Vectors Carrying Polylinker Sequences, *J. Mol. Biol.* **170,** 827–842.
6. Collins, J. (1979), *Escherichia coli* Plasmids Packageable *in Vitro* in Lambda Bacteriophage Particles, *Meth. Enzymol.* **68,** 309–326.
7. Maniatis, T., Fritsch, E. F., and Sambrook J. (1982), *Molecular Cloning,* Cold Spring Harbor Laboratory, New York.

# Chapter 50

# DNA Transformation of Mammalian Cells

***Jeffrey W. Pollard, Yunus Luqmani, Alan Bateson, and Kokila Chotai***

*MRC Research Group in Human Genetic Diseases, Department of Biochemistry, Queen Elizabeth College, University of London, London, United Kingdom*

## Introduction

The manipulation of gene sequences between cells is a fundamental technique in genetics. Mammalian cells will take up and express genes when they are exposed to either metaphase chromosomes or naked genomic or recombinant DNA. In each case the uptake and expression is enhanced by the formation of a DNA–calcium phosphate precipitate (*1,2*). Alternatively, cloned recombinant DNA sequences may be introduced directly into mammalian cells by fusion with bacterial protoplasts containing recombinant plasmids (*3*) or by microinjection of purified

DNA directly into the cell's nucleus (*4*). In many cases the transforming DNA is stably expressed, and consequently alters the genotype of the recipient cell. Since long-term transformation is a relatively rare event, identification of this altered genotype requires the use of genes coding for selectable functions or for proteins easily assayable by single cell antibody or related techniques.

This alteration of genotype enables the investigation of the relationship between gene structure and function and also acts as an assay for specific gene sequences. Thus, providing there is a means of rapidly assaying a gene's function, DNA transformation in principle provides a means to clone any gene (*5*). DNA transformation also has a therapeutic potential, since it may eventually be possible to correct specific gene defects in, for example, fibroblasts in vitro, and reintroduce these reconstituted cells back into the patient.

This chapter will describe a single method to transfer metaphase chromosomes and the three most commonly used techniques for introducing DNA into mammalian cells.

## Materials

1. Complete Growth Medium: Alpha-minimal essential medium plus 10% (v/v) fetal calf serum, but any tissue culture medium will suffice.
2. Tris-$CaCl_2$ Buffer: 15 m*M* Tris-HCl, pH 7.0, and 3 m*M* $CaCl_2$.
3. HBS Buffer: 140 m*M* NaCl, 25 m*M* Hepes, and 0.75 m*M* $Na_2HPO_4$, pH 7.12, at 20°C. The pH is critical.
4. TEN Buffer: 10 m*M* Tris-HCl, pH 7.9, at 20°C, 10 m*M* EDTA, and 10 m*M* NaCl.
5. TE Buffer: 1 m*M* Tris-HCl, pH 7.9, at 20°C, 0.1 m*M* EDTA.
6. PBS: 0.14*M* NaCl, 2.7 m*M* KCl, 1.5 m*M* $KH_2PO_4$, 8.1 m*M* $Na_2HPO_4$.
7. Pronase: Prepared by dissolving 25.5 mg in 5.1 mL of water and heating it to 56°C for 15 min followed by a 1 h incubation at 37°C. Store at −20°C.

8. L-Broth: 10 g Bacto-tryptone, 5 g Bacto-yeast extract, and 10 g NaCl adjusted to pH 7.5 with NaOH and autoclaved.
9. Chloramphenicol: 34 mg/mL in ethanol. Stored at −20°C.
10. Lysozyme: 50 mg/mL in water. This should be dispensed into aliquots and stored at −20°C. Discard each aliquot after use; do not refreeze.
11. PEG 1000: 42–50% polyethylene glycol solutions are prepared in Tris-buffered saline (*12*) by heating the PEG to 65°C and dispensing the appropriate volume. This is autoclaved, and when cooled to 50°C, sterile Tris-buffered saline is added.
12. Tris-Buffered Saline: 10 m*M* Tris-HCl, pH 7.4, and 0.15*M* NaCl.
13. Kanamycin (25 mg/mL) and nalidixic acid (20 mg/mL) in water. This is sterilized by filtration through a 2 μm Millipore filter and stored in aliquots at −20°C.
14. Dimethyldichlorosilane solution, used as supplied commercially.

All solutions must be sterile and unless otherwise specified are stored at 4°C.

# Method

## *Chromosome-Mediated Gene Transfer*

1. Cells are laid down in 10–20 75 mL flasks at $2 \times 10^6$ cells/flask in complete growth medium, 2 d prior to the experiment.
2. At 16 h before the experiment, colcemid (0.06 μg/mL) is added to each flask to effect mitotic arrest. At the end of this incubation mitotic cells are harvested by selective detachment from the monolayer by gentle shaking or, alternatively, all the cells are collected by trypsinization (0.1% trypsin in PBS-citrate).
3. Cells are harvested by centrifugation at $400g_{av}$ for 3.5 min and resuspended in 50 mL of cold hypotonic, 75 m*M* KCl. After a 10 min incubation at 4°C, the cells are

collected by centrifugation at $300g_{av}$ for 5 min and resuspended in Tris–$CaCl_2$ buffer at about $2 \times 10^6$ cells/mL.

4. Triton X-100 is added to 1% and the cell suspension homogenized with six strokes of a Dounce homogenizer at 4°C and the homogenate diluted by a factor of two.
5. The cell debris is removed by centrifugation at $50g_{av}$ for 7 min and the supernatant retained. The pellet is resuspended in Tris–$CaCl_2$ buffer and recentrifuged at $50g_{av}$ for 7 min and this procedure repeated twice. The supernatants are pooled and the chromosome number determined by counting in a hemocytometer with a phase contrast microscope.
6. The chromosomes are collected by centrifugation for 25 min at $800g_{av}$ at 4°C. The pellets are resuspended in Tris–$CaCl_2$ buffer and recentrifuged at $800g_{av}$ for 30 min.
7. The final washed pellet of chromosomes is resuspended in 1 × HBS at about 2–4 × $10^6$ chromosome cell equivalents/mL. The chromosomes are now ready for transfer.
8. The recipient cells should be growing exponentially in complete growth medium in 75 mL flasks at about $2 \times 10^6$ cells/flask on the day of experimentation. Immediately before the chromosome transfer, the medium is replaced with fresh, warm complete growth medium.
9. To the chromosome preparation in HBS buffer at room temperature, 2.5*M* $CaCl_2$ is added to a final concentration of 125 m*M* with bubbling (as described on page 326, 10 and 11).
10. The calcium phosphate precipitate is allowed to develop for 30 min, gently sheared with a plastic pipet, and 2 mL added to each flask.
11. After a 20-h absorption period at 37°C in a $CO_2$ buffered incubator, 10% (v/v) DMSO is added, being careful to mix the DMSO with the medium before exposure to the cells. Continue the DMSO treatment for 30 min (although some workers have recommended 50% DMSO for 2 min), aspirate the medium, and replace it with 20 mL of fresh growth medium.

12. Thereafter leave the cells for an appropriate expression time of at least 18 h, but up to three cell generations.
13. Finally, trypsinize the cells and plate them at $5 \times 10^5$ cells/60 mm plate and put them in the appropriate selective conditions.

## *DNA Transformations*

### *Calcium Phosphate Technique*

For use as carrier DNA almost any method of DNA preparation that results in DNA of greater than 50 kb is suitable. Alternatively salmon sperm DNA may be purchased commercially. A quick method for preparing relatively impure DNA that works well as carrier DNA in gene transfer experiments is described in steps 1–7. Steps 8–12 describe the DNA transfer.

1. At 3 d prior to isolation, set up three 150T flasks with the appropriate cells plated at about one-tenth of their saturation density.
2. After 3 d, trypsinize the cells, neutralize the trypsin with 10% (v/v) serum, and wash the cells three times with 20 mL PBS.
3. Resuspend the cells in 5 mL TEN buffer and leave for 15 min at 4°C.
4. Add Triton X-100 to 1% from a 50-fold concentrated stock. Vortex hard and centrifuge at full speed (about $3000g_{max}$) in a refrigerated bench centrifuge for 5 min.
5. Pour off the supernatant, resuspend the pellet in 5 mL of TEN buffer, and recentrifuge without Triton.
6. Resuspend the pellet in 5 mL of TEN buffer and add 0.45 mL of a 5% (w/v) solution of SDS. Add pronase at 250 μg/mL (0.27–5.45 mL DNA). Digest at 37°C for 9 h followed by dialysis against 2 L of TE buffer at 4°C overnight. Change the dialysis buffer and continue dialysis for a further 8 h.
7. Transfer the resultant DNA solution to a sterile plastic tube and test 0.1 mL in tissue culture medium for sterility.

8. The recipient cells are laid into a 100 mm tissue culture dish at $5 \times 10^5$ cells in 10 mL of complete growth medium for 24 h before exposure to the DNA.
9. Set up two sterile tubes: the first contains DNA in TE buffer at 10–20 μg/mL final concentration and to this 1*M* $CaCl_2$ is added to 0.25*M*. If recombinant plasmid DNA is to be used, add at 1 ng–1 mg (usually 10–20 ng is all that is necessary). To the second tube add an equal volume of 2× strength HBS.
10. Into the second tube, the DNA is added slowly with gentle bubbling. This is achieved by attaching a plugged pipet to an electronic pipet aid and slowly bubbling air at about 1 bubble/s.
11. Leave the precipitate to develop for 45 min, gently pipet up and down about five times with a 5 mL plastic pipet, and add 1 mL of the resuspended pellet to each 100 mm tissue culture plate containing 10 mL of fresh medium.
12. Incubate overnight at 37°C followed by a change of medium. Incubate the cells for a further 24 h and then place them in the appropriate selective conditions.

## *Protoplast Fusion Technique*

1. Seed 100 mL of L-broth with 2 mL of an overnight bacterial starter culture and grow at 37°C until an $OD_{600}$ of 0.8 (approximately $6 \times 10^8$ cells/mL) is reached.
2. Amplify the plasmid sequence number by adding chloramphenicol to 170 μg/mL. Leave the culture shaking for 16 h.
3. Harvest the cells in 50 mL disposable plastic tubes at $3000g_{max}$ for 15 min.
4. Combine the pellets and wash twice with 50 mL of 10 m*M* Tris-HCl, pH 8.0 at 20°C.
5. Regain the cells by centrifugation and resuspend the pellet in 20% (w/v) sucrose in 0.1*M* Tris-HCl, pH 8.0, to give 10 OD units/mL (about 15 mL in total).
6. Transfer this to a 100 mL conical flask and warm to 37°C in a water bath with gentle stirring.
7. Add lysozyme to 100 μg/mL and continue to stir for 12 min. Remove from the water bath and over a period of 3 min add dropwise, while stirring, 1 vol 0.1*M* $Na_2EDTA$ to 10 vol cells. Be careful, since too rapid dilution results in cell lysis. Return to 37°C and visual-

ize the protoplasts by phase microscopy after a further 10 min. This completes the preparation of protoplasts.

8. To carry out protoplast fusion, 60 mm Petri dishes containing 5 mL complete growth medium are seeded 24 h prior to fusion to give 2–4 × $10^5$ cells/plate on the day of fusion.
9. The freshly prepared protoplasts are diluted with 3 vol of tissue culture medium lacking serum but containing 7 %(w/v) sucrose, 10 m*M* $MgCl_2$, and 0.2 μg DNase I/mL and incubated for 10 min.
10. The medium is removed from the cells and 5 mL of protoplast suspension added.
11. The protoplasts are centrifuged onto the monolayer in the 1 L swing out buckets of an MSE 6L centrifuge at 3000 rpm for 5 min at room temperature. After centrifugation the clear supernatant is removed carefully so as to avoid disturbing the protoplast layer.
12. To the cell/protoplast layer, add dropwise 1 mL of PEG 1000 rocking the plate to ensure it completely covers the cells. After 60–90 s add 5 mL of Tris-buffered saline and quickly aspirate. Wash the cells rapidly three times with 4 mL of Tris buffered saline. Add 5 mL of complete growth medium containing 100 μg/mL kanamycin and 50 μg/mL nalidixic acid.
13. Thereafter the cells are manipulated as required.

## *Microinjection Technique*

Microinjection is performed under phase contrast microscopy with a glass microcapillary prefilled from the tip with the DNA sample that is then directed into the cell by the use of a micromanipulator. The sample is ejected by air pressure exerted by a syringe connected to the capillary. Cells may be localized by griding coverslips and the biological expression of the DNA estimated by autoradiography, immunofluorescence, or growth in selective medium.

1. As an initial step, suitable needles have to be prepared. Glass tubes of 1.2 mm inner and 1.5 mm outer diameter are cut into 20 cm lengths and cleaned by placing in a glass cylinder in chromic acid for 30 min.

2. They are washed extensively in tap water and then placed in a solution of HCl (pH 3–4) for 30 min. Thereafter they are washed again extensively in tap water, followed by further washings with distilled and double-distilled water.
3. They are dried in an oven and siliconized. This is performed by dipping them in dimethyldichlorosilane solution, taking care to ensure that all air bubbles are removed.
4. After dipping, the tubes are washed with tap water and neutralized with a 2% solution of $NH_3$. Rinse again in tap water, distilled and double-distilled water, and again air dry. Store in a closed container until used.
5. The needle is first constricted to about 0.3 mm in diameter by heating the center with a small bunsen flame and gently pulling when the glass softens.
6. The glass tube is then placed in a needle puller (E. Leitz, West Germany) and, by melting the central part of the constriction with a wire heating forge and applying force to the end, it is pulled to the appropriate thickness. If conditions have been optimized, this will give needles about 0.1–0.5 μm inner diameter. To facilitate injecting into fibroblast nuclei, tips may be bent with the forge to 40° angle about 5 mm from the tip. Needles are stored upright in a suitably machined block of Perspex under a bell jar and sterilized by autoclaving.
7. To carry out microinjection, about 2000 fibroblasts are seeded onto cover slips the day prior to injection and incubated overnight in complete growth medium at 37°C.
8. The DNA sample (usually recombinant plasmid DNA) is dissolved either in an isotonic buffer (0.048*M* $K_2HPO_4$, 0.14*M* $NaH_2PO_4$, 0.0045*M* $KH_2PO_4$, pH 7.2) or TE buffer at concentrations up to 1 mg/mL and centrifuged at 10,000*g* for 10 min immediately before injection, when a drop is placed on a clean microscope slide.
9. The syringe is placed into the micromanipulator (E. Leitz, West Germany), set up on a solid block of concrete to reduce vibrations, and the tip is focused un-

der the microscope and introduced into the drop of sample, which is then sucked into the needle by capillary action and negative pressure from the syringe.

10. The needle is carefully moved out of the way and the open 60 mm dish containing the cover slips is placed on the microscope stage.
11. The needle is lowered almost into focus and individual cells approached by operating the micromanipulator controls with the right hand. A dent is observed on the cell when the needle touches and the injection, effected by pressure on the syringe with the left hand, is marked by a change in contrast of the nucleus. Do beware of damaging the cell by injecting too large a volume. After injection, the needle is moved onto the next cell. Using this technique, with practice, about 1000 cells/h can be injected and about $10^{-11}$ mL introduced per nucleus.
12. Thereafter, the cells are treated as required by the desired analytical technique.
13. Sterility is maintained under these conditions by performing the microinjection in an enclosed space that has been sterilized by UV germicidal lamps.

# Notes

1. The major problem for these techniques is to ensure that the recipient cell surviving after marker selection is genuinely transformed and is not simply the consequence of genetic reversion or, in the case of chromosome transfer, caused by contaminating cells carried over to the chromosome preparation. One test for this is to look for cells that unstably express the marker under nonselective conditions. This, coupled with an increased frequency of reversion over the controls in the original transformation, is a sure sign that the cells are transformants. Stable transformants also occur, however, and often these are the most useful cell types. It is, if possible, therefore advisable to construct experiments so that the recipient cell has a different and identifiable genetic background from the donor cells. Similarly, if recombinant DNA is transferred, total genomic

DNA of the transformant may be probed by Southern gel analysis (Chapter 9) to show integration of foreign DNA sequences into high molecular weight DNA or, if interspecific DNA transfer has been performed, to probe with cloned repeat sequences from the donor species. Also, in some cases the cloning vector may have a gene that is dominantly selectable in mammalian cells; thus resistance by the transformants indicates that transfer at least of the vector has taken place. Nevertheless, in many cases a simple increase in reversion frequency is taken as evidence that transfer has occurred. Under these circumstances, the investigator must exercise caution in interpretation of the results and adequate controls must be built into the experimental design. Microinjection avoids many of these problems since individual cells may be identified, but as yet only recombinant DNA has been introduced by this technique; genomic DNA being too viscous to pass through the needle without substantial shearing.

2. Technically, with the exception of microinjection, these DNA transfer methods are simple and, providing adequate sterility control is maintained, should present few problems. The major variability in the technique is the transformation efficiencies of different cell lines. Thus Chinese Hamster Ovary cells are 100 time less efficient in the stable expression of exogenous DNA than mouse L cells. The basis of this variability has yet to be understood.
3. The calcium phosphate technique can result in a low frequency of transfer if the precipitate is not of the right texture. Photographs of the rather coarse precipitate may be found in Graham et al. (2). For DNA transformations, carrier DNA size and concentration also affects the transformation efficiency. The DNA should be 50–60 kb at a final concentration of 10 μg/mL. Treatment with butyrate may also increase this frequency in a cell-dependent manner (*8*).
4. Transformation efficiencies of one colony in $10^{-4}$ may be achieved with chromosome transfer and one in $10^{-5}$ or $10^{-6}$ with total genomic DNA. Using cloned

recombinant DNA transformation efficiencies of 100 colonies/ng DNA may be obtained with a maximum of 1–2 transformed colonies/$10^3$ recipient cells. Up to 80% of the cells, however, may transiently express the exogenous DNA.

5. Protoplast fusion can only be used for transfer of cloned DNA sequences. The critical factors in this technique are the formation of protoplasts and the length of exposure to PEG. Unfortunately, we have found that every strain of *E. coli* is different in the speed and efficiency of protoplast formation and that commonly used strains such as HB101 are rather poor in protoplast formation, requiring a longer treatment time with lysozyme. This side of the technique needs to be optimized for each strain (*7*). In all cases protoplasts should be used immediately after preparation. Chloramphenicol amplification of the plasmid sequences is also essential for successful transfer.
6. PEG is toxic to cells and this toxicity varies for each cell type; again this needs to be optimized for a particular experimental situation. Removal of $Ca^{2+}$ or addition of DMSO has been reported to reduce PEG toxicity. Cells should also be subconfluent to reduce cell to cell fusion.
7. Protoplast fusions can result in 50–100% of the cells showing transient expression of the donor gene, with at best one colony per 500 cells following outgrowth in selective medium. Caution must be excercised with this technique since during selection against auxotrophic mutants the residual bacterial protoplasts can metabolize sufficient substrate to cause break-through in the selection.
8. Microinjection requires sophisticated equipment and therefore is not as useful as the other techniques and it is also best learned by demonstration. Nevertheless, once the equipment is purchased it is a simple technique that has yielded valuable information (*4,6*). Capecchi (*4*) has reported that up to 100% of the injected cells express the donor gene and that one cell in 500–1000 will give colonies following outgrowth in selective conditions.

## Acknowledgments

This article is based on DNA transfer studies supported by grants from the MRC to JWP and by an EMBO short term fellowship to JWP in Dr. J. Celis' laboratory in Aarhus, Denmark. AB is a MRC research student.

## References

1. Lewis, W. H., Srinivasan, P. R., Stokoe, N., and Siminovitch, L. (1980) Parameters governing the transfer of genes for thymidine kinase and dihydrofolate reductase into mouse cells using metaphase chromosomes or DNA. *Somat. Cell Genet.* **6,** 333–347.
2. Graham, F. L., Bacchetti, S., McKinnon, R., Stanners, C. P., Cordell, B., and Goodman, H. M. (1980) Transformation of mammalian cells with DNA using the calcium technique, in *Introduction of Macromolecules into viable Mammalian Cells* (Baserga, R., Croce, C., and Rovera, G., eds.) Winter Symposium Series **1,** pp. 3–25, Alan R. Liss, New York.
3. Schaffner, W. (1980) Direct transfer of cloned genes from bacteria to mammalian cells. *Proc. Natl. Acad. Sci. USA* **77,** 2163–2167.
4. Capecchi, M. R. (1980) High efficiency transformation by direct microinjection of DNA into cultured mammalian cells. *Cell* **22,** 479–488.
5. Pellicer, A., Robins, D., Wold, B., Sweet, R., Jackson, J., Lowy, I., Roberts, J. M., Sim, G. K., Silverstein, S., and Axel, R. (1980) Altering genotype and phenotype by DNA-mediated gene transfer. *Science* **209,** 1414–1422.
6. Graessmann, A., Graessmann, M., and Mueller, C. (1980) Microinjection of early SV40 DNA fragments and T-antigen. *Meth. Enzymology* **65,** 74–83.
7. Sandri-Goldin, R. M., Goldin, A. L., Levine, M., and Glorioso, J. (1983) High efficiency transfer of DNA into eukaryotic cells by protoplast fusion. *Meth. Enzymology,* **101,** 402–411.
8. Gorman, C. M., and Howard, B. H. (1983) Expression of recombinant plasmids in mammalian cells is enhanced by sodium butyrate. *Nucleic Acid Research* **11,** 7631–7648.

# Chapter 51

# Chemical Cleavage (Maxam and Gilbert) Method for DNA Sequence Determination

***Wim Gaastra***

*Department of Microbiology, The Technical University of Denmark, Lyngby, Denmark*

## Introduction

Unlike the determination of the amino acid sequence of proteins and peptides, which is based on the sequential degradation of these structures and the subsequent identification of the cleaved-off amino acid residue, DNA sequence analysis is based on the high-resolution electrophoresis on denaturing polyacrylamide gels of oligonucleotides with one common end and varying in length by a single nucleotide at the other end (Chapter 52). In the Maxam and Gilbert method for DNA sequencing (*1,2*), the four sets of oligonucleotides are obtained by treating a $^{32}$P-end-labeled DNA fragment (Chapters 39–41) under four different conditions with a reagent that modifies a particular nucleotide, followed by cleavage of the DNA molecule next to the modified nucleotide. It is for this reason that the method is also known as the partial chemical degradation method for DNA sequencing. The other, and probably most frequently used method for sequencing DNA, mainly developed by Sanger and his colleagues (*3,4*) is described in Chapter 53. In the Maxam and Gilbert method, the reaction conditions are chosen in such a way that only a limited number of cleavages occurs in each DNA molecule. Ideally, they should be such that only one nucleotide in each labeled molecule inside the region to be sequenced is modified. Since any nucleotide is as likely to react as any other with the reagent that is specific for it, the one-hit reactions evenly distribute the radioactivity among the cleavage products.

Modification of the four nucleotides is obtained in the following ways. Guanine is methylated with dimethyl sulfate, adenine is methylated by formic acid. Hydrazine reacts with the pyrimidines, cytosine and thymine, removes the bases, and leaves a hydrazone. In the presence of a high concentration of sodium chloride, the reaction with thymine residues is preferentially suppressed. After these reactions, the four samples are incubated with piperidine, which cleaves the sugar phosphate backbone of the DNA molecule where this is modified. The fragments so produced are then analyzed by gel electrophoresis (*See* Chapter 52).

## Materials

1. Cacodylate buffer: 50 m*M* sodium cacodylate; 10 m*M* magnesium chloride; 0.1 m*M* EDTA, pH 7.4.
2. Dimethyl sulfate (DMS) (99%).
3. DMS stop mixture: 1.0*M* Tris-acetate; 1.5*M* sodium acetate; 1.0*M* β-mercaptoethanol; 50 m*M* magnesium acetate; 1 m*M* EDTA; 0.4 mg/mL tRNA; pH 7.5.
4. Formic acid, 100%.
5. Acetate RNA mix: 0.3*M* sodium acetate; 0.1 m*M* EDTA; 0.4 mg/mL t-RNA.
6. Hydrazine, 96%.
7. Sodium chloride, 5*M*.

8. Ethanol, 96%.
9. Sodium acetate, 0.3*M*.
10. Ethanol, 80%.
11. Piperidine, 99%.
12. Formamide (deionized) dye solution: 1 mL formamide; 20 μL 1% Orange G; 10 μL 0.1% Bromophenol Blue; 20 μL 1% *p*-Xylenol Blue.

# Method

1. Dissolve the end-labeled DNA fragment in 30 μL of $H_2O$.
2. Divide this solution between four Eppendorf tubes for the four different reactions. Take 5 μL of DNA solution for the guanine- and cytosine-specific reactions, and 10 μL of DNA for the adenine + guanine and thymine + cytosine reactions, respectively. Keep the tubes at 0°C on ice.

## *The Guanine Reaction*

1. Add 200 μL of cacodylate buffer (at 0°C), and 2 μL of DMS, then mix and incubate for 4.5 min at 20°C.
2. After this time, add 50 μL of DMS stop mixture (at 0°C), mix, and chill in icewater.
3. Add 750 μL of 96% ethanol (at −20°C), mix, and precipitate the DNA for 5 min at −70°C, then centrifuge for 5 min in an Eppendorf centrifuge. Discard the supernatant (*see* Note 4).
4. Dissolve the pellet in 200 μL of 0.3*M* sodium acetate, add 600 μL of 96% ethanol, then mix and precipitate the DNA for 5 min at −70°C.
5. Centrifuge for 5 min in an Eppendorf centrifuge, discard the supernatant, then wash the pellet in 1 mL of 80% ethanol.
6. Centrifuge for 5 min in an Eppendorf centrifuge, discard the supernatant, then wash the pellet in 1 mL of 96% ethanol.
7. Centrifuge for 5 min in an Eppendorf centrifuge, discard the supernatant, then dry the pellet *in vacuo*.

8. Dissolve the dry DNA pellet in 100 μL of a 10% piperidine solution (freshly made) and incubate at 95°C for 30 min in a water bath.
9. Lyophilize the DNA solution until a dry pellet is obtained, then dissolve the pellet in 50 μL of $H_2O$.
10. Lyophilize the DNA solution until a dry pellet is obtained, then dissolve the pellet in 10 μL of $H_2O$ and lyophilize the DNA solution until a dry pellet is obtained.
11. Dissolve the pellet in the formamide dye solution. The amount of formamide dye solution depends on the number of sequence gels one intends to run and on the amount of radioactivity present in the sample. Usually 1–2 μL are loaded, which should contain 10,000–20,000 cpm measured by Cerenkov counting, to ensure an exposure time of 1–2 d.

## *Adenine Plus Guanine Reaction*

1. Add 22 μL of 100% formic acid, mix, and incubate for 10 min at 20°C.
2. Add 200 μL of acetate–RNA mix (at 0°C), chill in ice water, then continue with steps 3–11 of the guanine reaction.

## *Cytosine Plus Thymine Reaction*

1. Add 10 μL of $H_2O$ (at 0°C) and then 30 μL of 96% hydrazine. Mix and incubate for 7 min at 20°C.
2. After this time, add 200 μL of acetate–RNA mix (at 0°C), chill in ice water, then continue with steps 3–11 of the guanine reaction.

## *Cytosine Reaction*

1. Add 15 μL of 5*M* NaCl and then 30 μL of 96% hydrazine. Mix and incubate for 10 min at 20°C.
2. After this time, add 200 μL of acetate–RNA mix (at 0°C), chill in ice water, then continue with steps 3–11 of the guanine reaction.

# Notes

1. In many protocols, the addition of carrier DNA to the various base-specific reaction mixtures is advised. If this is considered necessary, 1 µL of a 1 mg/mL solution of commercially available calf thymus DNA should be added to each of the four tubes prior to carrying out the chemical cleavage steps. The carrier DNA should be sheared and deproteinized.

   The procedure for preparing carrier DNA is as follows:

   (1) Dissolve 200 mg of calf thymus DNA in 2 mL of TE buffer and sonicate for 10 × 15 s (Branson Sonifier B12, Setting 4).
   (2) Extract the sample with phenol (3 × 2 mL) and then with ether (2 × 5 mL).
   (3) Dialyze the sample against distilled water for 18 h. Measure the absorbance at 260 nm and dilute the preparation to give a concentration of 2 mg/mL.
   (4) Divide the solution into 100 µL aliquots and store frozen.
2. The use of siliconized Eppendorf tubes is also advised in most protocols, but has not been found to be essential in our sequencing experiments.
3. Always put the Eppendorf tubes in the same orientation in the centrifuge, i.e., with the plastic connection of the tube and cap to the outside. In this way one always knows where the pellet is after the centrifugation.
4. Dimethyl sulfate is volatile and toxic and should only be handled in a fume hood with gloves on. It can be inactivated by reaction with alkali. The supernatant obtained in step 3 of the guanine reaction should therefore be discarded in 5*M* sodium hydroxide.
   **N.B.:** This should not be done with large amounts of DMS.
5. Hydrazine is toxic and in the anhydrous state explosive. The reagent should be handled with care and al-

Table 1
Flowsheet for the Maxam and Gilbert Sequencing Method

| Specific reaction | G | G + A | C + T | C | A + C |
|---|---|---|---|---|---|
| Chill in ice water | 5 µL DNA<br>1 µL ct DNA | 10 µL DNA<br>1 µL ct DNA | 10 µL DNA<br>1 µL ct DNA | 5 µL DNA<br>1 µL ct DNA | 10 µL DNA<br>1 µL ct DNA |
| Add and mix | 200 µL cacodylate buffer | 22 µL formic acid | 15 µL $H_2O$ | 15 µL 5*M* NaCl | 100 µL 1,2 *N* NaOH/EDTA 1 m*M* |
| | 2 µL DMS | 22 µL formic acid | 30 µL Hydrazine | 30 µL Hydrazine | |
| Incubate | 4.5 min 20°C | 10 min 20°C | 7 min 20°C | 10 min 20°C | 10 min 90°C |
| Add, mix, and chill | 50 µL DMS stop mixture (of 0°C) | 200 µL acetate tRNA mixture (of 0°C) | 200 µL acetate tRNA mixture (of 0°C) | 200 µL acetate tRNA mixture (of 0°C) | 150 µL 1*N* acetic acid + 1 µL tRNA |
| Add as soon as possible after mixing | 750 µL ethanol (96%) (−20°C) | See G | See G | See G | See G |
| Precipitate | 5 min at −70°C | See G | See G | See G | See G |

| | | | | | |
|---|---|---|---|---|---|
| Centrifuge | 5 min | See G | See G | See G | See G |
| | Discard supernatant | See G | See G | See G | See G |
| Add and mix | 200 µL 0.3*M* | See G | See G | See G | See G |
| | Sodium acetate | See G | See G | See G | See G |
| | 600 µL ethanol (96%) | See G | See G | See G | See G |
| Precipitate | 5 min at −70°C | See G | See G | See G | See G |
| Centrifuge | 5 min | See G | See G | See G | See G |
| | Discard supernatant | See G | See G | See G | See G |
| Add and wash with | 1 mL 80% ethanol | See G | See G | See G | See G |
| Centrigfuge | 5 min | See G | See G | See G | See G |
| Add and wash with | 1 mL 96% ethanol | See G | See G | See G | See G |
| Centrifuge | 5 min | See G | See G | See G | See G |
| Dry pellet and continue with point 8 of the protocol | | | | | |

ways be kept in a fume hood. It can be inactivated by reaction with ferric chloride. The supernatants obtained after step 2 of the cytosine plus thymine and the cytosine reactions should therefore be discarded into 3*M* ferric chloride.
**N.B.:** This should not be done with large amounts of hydrazine.

6. For the lyophilization after the cleavage reaction with piperidine, the use of a Speedvac concentrator (Savant) has proved to to be very convenient. Usually only one lyophilization is needed with this instrument, since the DNA is already concentrated in the bottom of the tubes by the centrifugal force.
7. The pellets should be completely dry before adding the formamide dye mixture otherwise the fragments may show anomalous mobility upon subsequent electrophoresis.
8. The amount of DNA in the adenine plus guanine reaction and in the cytidine plus thymine reaction is doubled to assure even labeling of all bands since in these cases roughly double the amount of bands is obtained.
9. Care should be taken that the number of samples to be sequenced does not exceed the number of holes in the Eppendorf centrifuge. This considerably slows down the procedure, which can lead to modification of more bases than intended.
10. If very long sequences are to be read, i.e., on 1-m-long gels the reaction times for the cleavage steps given in the protocols should be halved or even shortened more.
11. Steps 2–5 of each protocol should be performed with as little loss of time as possible, to make sure that the reagents are removed and inactivated as soon as possible after the reactions have been stopped.
12. The use of a flow sheet as given in Table 1 is recommended.
13. Adenine rings, but also cytosine rings, can be opened adjacent to the glycosidic bond by strong alkali. The opened bases can also be displaced by piperidine and 3′- and 5′-phosphates are readily eliminated from this product, which leads to strand scission wherever the

rings were originally opened. The reaction is about 85% adenine specific and 15% cytosine specific and is included in Table 1.

14. Piperidine can readily be oxidized and should come in contact with air as little as possible. The 10% piperidine solution should therefore always be freshly made shortly before use.
15. The tRNA present in some solutions is simply there to act as a carrier at the DNA precipitation steps.

## References


1. Maxam, A. M., and Gilbert, W. (1977) A new method for sequencing DNA. *Proc. Natl. Acad. Sci. USA* **74,** 560–564.
2. Maxam, A. M., and Gilbert, W. (1980) Sequencing end labelled DNA with base-specific chemical cleavages, *in* Grossman, L., and Moldave, K. (eds.), *Meth. Enzymol.* **65,** 499–560.
3. Sanger, F., and Coulson, A. R. (1975) A rapid method for determining sequences in DNA by primed synthesis with DNA polymerase. *J. Mol. Biol.* **94,** 441–448.
4. Sanger, F., Nicklen, S., and Coulson, A. R. (1977). DNA sequencing with chain terminating inhibitors. *Proc. Natl. Acad. Sci. USA* **74,** 5463–5467.

# Chapter 52

# Gel Electrophoretic Analysis of DNA Sequencing Products

***Wim Gaastra***

*Department of Microbiology, The Technical University of Denmark, Lyngby, Denmark*

## Introduction

As mentioned in Chapter 51, DNA sequence analysis is based on high-resolution electrophoresis on denaturing polyacrylamide gels. In both the partial chemical cleavage method (Chapter 51) and the partial resynthesis method (Chapter 53), the labeled DNA fragments to be separated have one common end and the other end varies in length. The minimal size difference of DNA fragments in adjacent tracks of the gel is one nucleotide. The resolution of the gel should therefore be high enough to see two DNA molecules differing in size by one nucleotide as two separate bands. The resolution of the gel can be increased by decreasing the thickness of the gel and by decreasing the volume of the sample applied to each slot. Several percentages of polyacrylamide are routinely used for DNA sequence gels (i.e., 6, 8, 12, and 20% acrylamide) depending on which part of the sequence is to be read. The higher acrylamide percentage gels are used to determine the sequence of the first 50–100 nucleotides. The others are used to provide sequence data that overlaps the sequence obtained on the high percentage gel, and extends the sequence up to 200–250 nucleotides. Very long sequences can be read if the size of the gel is increased considerably, e.g., to 1 m. On these gels up to 400 nucleotides can be read. The gels are run under conditions that assure a temperature between 50–60°C at the surface of the plates. This, together with the high concentration of urea in the gel, is to make sure that the DNA does not renature during electrophoresis. The DNA molecules are denatured before loading and should not renature in order to obtain the correct distance between bands differing in size by only one nucleotide.

Several types of apparatus are commercially available now, but gel apparatus for DNA sequencing is still often homemade. This results in large differences in the sizes of the gels between various laboratories. In general, however, the gels are about 40 cm high and 0.3 mm thick.

## Materials

1. Tris-borate buffer (10× TBE): 0.9*M* Tris/HCl; 0.9*M* boric acid; 25 m*M* EDTA, pH 8.3.
2. Acrylamide stock solution (40% w/v): 38% (w/v) acrylamide, 2% (w/v) *N,N'*-methylene bisacrylamide The solution should be deionized by stirring it for 2 h with Mischbett-V ion exchange resin (Merck) or Amberlite $MB_1$ (BDH). The ion exchange resin is removed by filtration.
3. Ultrapure grade urea.
4. Ammonium peroxodisulfate solution in water, 10% (w/v)
5. *N,N,N'N'*-Tetramethylethylenediamine (TEMED).
6. Repelcoat (BDH) (optional, see Note 2).
7. PVC tape (or silicon grease).
8. Plastic syringe, 50 mL.
9. Slot formers, side spacers, and glass plates of desired dimensions. The glass plates are usually 20 × 40 cm

(in our institute 30 × 40 cm) or 20 × 80 cm when very long sequences are to be read (in our institute 30 × 80 cm). One of the plates should have a notch cut out at the top end (18 × 2 cm in a 20 cm wide plate) but it is also possible to use a plate that is 2 cm shorter than the first plate and assure leakproof mounting of the plates in the electrophoresis apparatus with modeling wax. The spacers are 0.5 × 40 cm and 0.3–0.5 mm thick, depending on the material available, It is clear that it would be easier to pour a gel when it is thicker, but this leads to loss of resolution. Spacers and slot formers are cut out of Plasticard or Teflon. The slot former contains slots with teeth 10 mm deep, 4 mm wide, and 2 mm apart. The number of slots depends of course on the width of the plates used.

10. Hamilton syringe or glass capillaries, 10 μL.

# Method

1. Thoroughly clean two glass plates, one with a notch and one without. The plates should first be cleaned with a detergent and care should be taken that all residues of acrylamide from previous runs are removed. The plates are then rinsed with water and acetone to remove any trace of fat or grease and finally with 96% ethanol.
2. Assemble the plates and spacers. Both sides and the bottom are sealed with strongly adhesive PVC tape and the plates are clamped with foldback clips. Alternatively, the spacers can be greased with silicon grease. A spacer for the bottom has to be used as well in this case. The plates are then clamped in the same way as above with foldback clips. The use of grease has the disadvantage that the cleaning procedure has to be even more thorough.
3. Prepare an acrylamide mixture according to the scheme given below, depending on which percentage of acrylamide should be used.
   Add distilled $H_2O$ to make the total volume to 60 mL, then place the solution in a vacuum desiccator and degas. About 50 mL of gel mixture should be enough for a gel of dimensions 20 × 40 × 0.3 mm.

| Gel percentage | 6% | 8% | 12% | 20% |
|---|---|---|---|---|
| 40% acrylamide stock solution, mL | 9 | 12 | 18 | 30 |
| Urea, g | 30 | 30 | 30 | 30 |
| 10 × TBE buffer, mL | 6 | 6 | 6 | 3 |
| 10% ammonium peroxodisulfate, μL | 420 | 420 | 420 | 420 |

4. After degassing, add 20 μL of TEMED and mix gently.
5. Cast the gel by pouring the gel mixture between the plates with a 50 mL syringe (*see* Note 3). The plates are held at an angle of about 45° to prevent air bubbles from being trapped. For the same reason, the gel mixture is poured down along the edge of the plates. If any air bubbles get trapped, gently tap the glass with a piece of rubber tubing to force the bubbles out. The plates are gradually lowered during filling until they are in a horizontal position and the slot former is placed in position.
6. The gel is then left to polymerize (usually within 10–30 min). The gel can either be used directly or left overnight if care is taken that it does not dry out. This can be done by wetting some tissue with TBE buffer, putting it on top of the slot former, and wrapping the gel in Saran wrap.
7. Before mounting the gel on the gel apparatus, the PVC tape or the spacer in the bottom of the gel is removed, together with the slot former. Great care should be taken to rinse out the slots. Any unpolymerized acrylamide as well as small pieces of polyacrylamide gel debris should be completely removed since they obstruct the loading of the samples.
8. Install the gel in the electrophoresis apparatus and fill the buffer reservoirs with 0.5 × TBE buffer in the case of a 20% polyacrylamide gel and with 1 × TBE buffer for the other gels.
9. If the gel apparatus is thermostated, start circulating water of 55°C and prerun the gel for 1 h at 30 W. Nonthermostated gels should also heat up during this pre-electrophoresis to about 55°C. This temperature, together with the urea present in the gel, is necessary to keep the DNA completely denatured.

10. Just prior to loading, the samples are heated for 3 min in a boiling water bath to obtain single-stranded DNA fragments. The tubes are chilled in ice water immediately after boiling to prevent the DNA from renaturing and the slots are flushed with TBE buffer to remove any urea that might have diffused from the gel. Failure to do this will make sample layering difficult.
11. Load 1–2 μL samples. The samples can be loaded with a 10 μL Hamilton syringe or with a drawnout capillary tube. The capillary should be drawn out in such a way that it is easy to enter it in the slot, but should not be so narrow that filling and discharging become difficult.
12. Run the gel at constant power (40 W, voltage between 1500 and 2000 V, current about 20 mA) for the 40 cm gel. The 80 cm gels are run at a constant power of 60 W (voltage between 3000 and 5000 V, current about 25 mA). The length of the run is determined by what part of the sequence one wants to read. (*See* Notes section for the mobility of the dye markers used in comparison with oligonucleotides.)
13. At the end of the run, carefully prise the plates apart. Care should be taken that the gel does not stick partially to one plate and partially to the other.
14. The gel is then wrapped in Saran wrap and the positions of the dye markers indicated with radioactive ink. Identification marks for samples and the application order are applied with the same ink.
15. The gel then is covered with an X-ray film and exposed in an exposure box in a −80°C freezer for the time necessary (*See* Note 8) to obtain a good image of the gel (*see* Chapter 8 for autoradiography).
16. It is then necessary to interpret the autoradiogram. If the four sequence ladders are applied in the correct order (G, A + G, C + T, C), the two inner tracks contain all the possible bands, which makes it easy to count the number of bases. The sequence is read from the bottom of the gel to the top. One starts with the band with the highest electrophoretic mobility and continues with the one with the next highest, etc. If

two bands run at the same position in the G and the G + A track, the residue is a G; a band only in the G + A track indicates an A. Similarly, two bands running at the same position in the C + T and the C tracks mean that the residue is a C, whereas a band only in the C + T track indicates a T.

Normally the sequence is first read on a 20% acrylamide gel until the bands are spaced so close together that they cannot be discriminated. This is usually after 50–80 bases .Then one continues reading the sequence of the same sample on an 8% acrylamide gel where one usually can pick up the sequence between base 30–40 (i.e., it overlaps the sequence obtained on the 20% gel). The sequence on this 8% gel can be read to about 200–250 bases. In the ideal case the bands are regularly spaced and of almost equal intensity; unfortunately this is never so in practice. Bands are found too close together, too far apart, too faint, too intense, or missing.

One of the main problems encountered in the partial chemical degradation method (*1*) is very weak thymine bands in the C + T track. Usually this is caused by insufficient removal of salt from ethanol precipitations. In a row of succeeding Gs the band can become fainter and fainter until at 4 or 5 Gs it seems to be an A.

As one proceeds further up the gel, the bands will generally get weaker, since considering the design of the chemical cleavage approach, there will necessarily be a greater number of small cleavage products than of the larger ones. As a consequence of this a hole in the sequence as one approaches the top of the gel usually means an A or a T since the cleavage method for both of these bases involves additional cleavage at a second base (*see* Chapter 51) and consequently the available radioactivity is reduced by half, and may therefore not be detected on a weak sample. If there is a possibility to form a stable secondary structure in the single-stranded DNA, the fragment can run faster than it should. This will give rise to a very dark band and a hole in the sequence.

It is clear that the interpretation of sequence gels is very subjective and it is therefore a good idea to have the sequence read by two different people and to confirm it by sequencing the complementary strand. A trouble-shooting list for the most common problems encountered is given in ref. *1*.

## Notes

1. In the protocol it is said that the acrylamide solution should be degassed before pouring the gel. In practice, this is not absolutely necessary. Eliminating the degassing step increases the time needed for polymerization, but does not interfere with the structure and the resolution of the gel.
2. In many laboratories the notched glass plate is siliconized with Repel-coat (BDH) on the inner surface to facilitate the removal of this plate after the electrophoresis. We generally do not do this.
3. While assembling the glass plates and taping them, it is a good idea to insert the needle of the 50 mL syringe at this early stage. Usually it is very difficult to insert the needle after the plates have been taped since there is only a space of 0.3 mm. Leaving the needle after pouring the gel makes it easier to insert the slot-former. After introduction of the slot-former, the needle is removed and the slot-former is clamped to prevent acrylamide from polymerizing between the slot-former and the glass plate, which might interfere with the loading of the sample. These clamps should, however, not exert as strong a pressure as the foldback clips used on the sides of the gel.
4. Having assembled the gel in the gel apparatus, care should be taken that no air bubbles remain at the bottom of the gel, since this will produce an uneven current flow through the gel.
5. While heating the samples prior to loading, increased air pressure into tubes can cause the caps to come off the Eppendorf tubes. We have therefore constructed a rack for those tubes that makes it possible to tightly

clamp a lid over the top of the tubes that prevents the caps from flipping open.

6. In a 20% acrylamide gel, the dye markers run as follows: Orange G in the position of an oligonucleotide of four residues. Bromophenol blue in the position of an oligonucleotide of 10 residues.
   In a 12% acrylamide gel, bromophenol blue runs in the same position as an oligonucleotide of 12 residues and xylene xyanol FF at 42 residues. In an 8% acrylamide gel, these figures are 20 and 72 residues, respectively.
7. As described in Volume 1 it is also possible to dry the gels after elecrophoresis. If this is done, the loss of resolution during exposure is minimized. Exposure of dried gels can also take place at room temperature.
8. Using a mini-monitor Geiger-Müller counter, type 5.10 (Mini Instruments), the following exposure times were found necessary:

| | |
|---|---|
| 5 cps/slot | 21 days |
| 10 cps/slot | 14 days |
| 20 cps/slot | 7 days |

## References


1. Maxam, A. M., and Gilbert, W. (1980) Sequencing End Labelled DNA with Base Specific Chemical Cleavages, in Grossman, L. and Moldave, K. (eds.), *Meth. Enzymol.* **65,** 499–560.
2. Sanger, F., and Coulson, A. R. (1978) The use of thin acrylamide gels for DNA sequencing. *FEBS Lett.* **87,** 107–110.
3. Maat, J., and Smith, A. J. H. (1978) A method for sequencing restriction fragments with dideoxynucleoside triphosphates. *Nucleic Acids Res.* **5,** 4537–4546.
4. Smith, A. J. H. (1980). DNA sequence analysis by primed synthesis, *in* Grossman, L., and Moldave, K. (eds.), *Meth. Enzymol.* **65,** 560–580.

# Chapter 53

# DNA Sequence Determination Using Dideoxy Analogs

*M. J. Owen*

*ICRF Tumor Immunology Unit, Department of Zoology, University College London, London, England*

## Introduction

Dideoxy chain termination DNA sequencing was developed by Sanger and colleagues (*1, 2*) and is a simple and extremely accurate method of obtaining thousands of bases of sequence data per day. The procedure requires a single-stranded DNA template and a primer complementary to the 3′ end of the region of DNA to be sequenced. In the presence of the four nucleoside triphosphates (dNTPs), one of which is radioactively labeled, and the Klenow fragment of DNA polymerase, primed synthesis occurs across the region of interest. In the presence of competing dideoxynucleoside triphosphates (ddNTPs), specific termination occurs at each of the four different nucleotides. Chain termination occurs when a ddNTP is incorporated randomly at the 3′ end of the growing chain; the chain cannot be further extended because the ddNTP lacks the 3′ hydroxyl group. If this is carried out separately with each of the ddNTPs and the four sets of reaction products are fractionated on a polyacrylamide gel, which separates the synthesized single strands according to length, the DNA sequence can be deduced from the ascending order of the bands in the four different tracks. The principle of this method is summarized in Fig. 1.

The single-stranded template DNA required by this procedure is provided by the use of the filamentous bacteriophage MI3 as a cloning vector (*3*). Several genetically engineered versions of M13 have been constructed (e.g., M13mp8, mp9, mp10, or mp11) to provide a variety of unique restriction sites in which to clone the DNA to be sequenced. Single-stranded DNA is secreted from the host into the culture medium from which it can be rapidly isolated. The primer most commonly used to initiate DNA synthesis is a synthetic heptadecadeoxyribonucleotide (17-mer), generally termed a "universal" primer since it can be used with any recombinant template (*4*).

The most widely applicable approach to DNA sequencing using the M13 system is to reduce the DNA to be sequenced into a random series of fragments prior to sequencing. This is the so-called "shotgun approach" (*5*). Since about 300 or more bases can be read from a sequencing gel, the length of the random fragments should be 300–600 base pairs (bp). By far the most elegant procedure for the generation of random fragments from large DNA molecules utilizes sonication (*5*). Other methods are DNase digestion (*6*) or, if a restriction map is known, digestion with restriction enzymes. It will be assumed that random fragments from the DNA molecule to be sequenced have been successfully cloned into M13. This chapter details the subsequent four steps involved in obtaining a DNA sequence; namely, preparation of the template, sequence reactions, polyacrylamide gel electrophoresis, and reading the sequence.

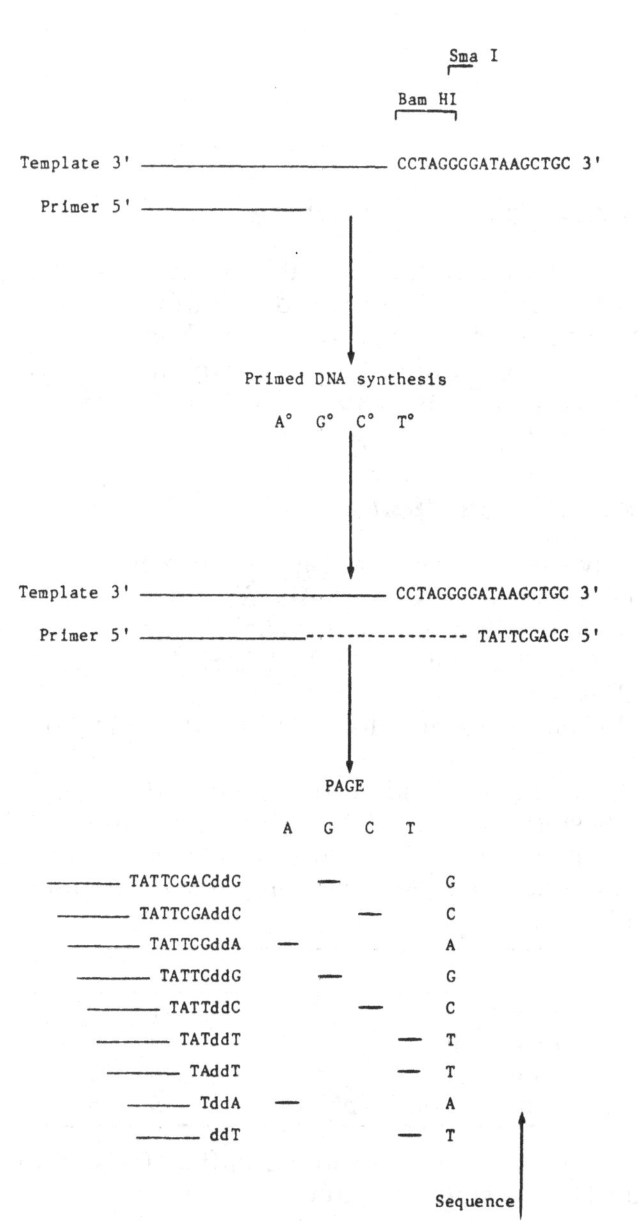


Fig. 1. Primed DNA synthesis is initiated by addition of the Klenow fragment of DNA polymerase I. In the example shown here, DNA has been cloned into the Sma site of M13mp8. DNA synthesis proceeds through the polylinker and into the insert. Only the sequence of the insert (TATT-) is shown. In the presence of the appropriate dideoxynucleotide (*see* Fig. 2 for the composition of A°, G°, C°, T°) chain termination occurs. These chains are separated according to length and the sequence read from the autoradiograph.

# Materials

## Preparation of the Template

1. M13 strains and JM101 are available commercially. JM101 should be restreaked monthly onto a minimal agar plate or maintained as a stock in glycerol.
2. 2 × TY media: Mix 10 g Bactotryptone, 10 g yeast extract, 5 g NaCl, and make 1 L with $H_2O$, pH to 7.4. Autoclave at 15 psi for 20 min.

## Sequencing Reactions

1. Universal sequencing primer (17-mer), 5'GTAAAACGACGGCCAGT3', and Klenow fragment of DNA polymerase are available commercially.
2. TE buffer contains 10 m*M* Tris-HCl, pH 8.0, 0.1 m*M* EDTA.
3. TM buffer contains 100 m*M* Tris-HCl, pH 8.0, 50 m*M* $MgCl_2$.
4. Deoxyribonucleoside triphosphates (dNTPs): Make 50 m*M* stock solutions in TE buffer. Solutions are stable indefinitely if stored at −20°C. Working solutions (dNTP mixs) have the following composition (volumes are in microliters):

| | T | C | G | A |
|---|---|---|---|---|
| 0.5 m*M* dT | 1 | 20 | 20 | 20 |
| 0.5 m*M* dC | 20 | 1 | 20 | 20 |
| 0.5 m*M* dG | 20 | 20 | 1 | 20 |
| 50 m*M* Tris-HCl, pH 8.0 | 5 | 5 | 5 | 5 |

5. Mix: make a solution in TE buffer 0.5 m*M* for each dNTP and store at −20°C.
6. Dideoxyribonucleoside triphosphates (ddNTTPs): make 10 *mM* stock solutions in TE buffer and store at −20°C.
7. Labeled deoxyadenosine triphosphate: [α-$^{32}$P]dATP is available commercially. The isotope should not be greater than 1 wk old.
8. Formamide dye mix: mix the following: 100 mL deionized formamide, 0.1 g xylene cyanol FF, 0.1 g

bromophenol blue, and 2 mL 0.5*M* EDTA. Store at room temperature.
To deionize formamide, stir with 2 g of Amberlite MB-1 (BDH) for 30 min and then filter.

## Polyacrylamide Gel Electrophoresis

1. Acrylamide gel solutions: 40% stock solution—make up 380 g acrylamide (2 × recrystallized) and 20 g bisacrylamide to 1 L with distilled $H_2O$. Deionize by stirring for 30 min with 20 g Amberlite MB-1; filter and store at 4°C.
   The 6% working solution: mix the following: 460 g urea (ultrapure), 100 mL 10 × TBE, 150 mL 40% acrylamide stock, and make to 1 L with distilled water. Store at 4°C.
2. 10 × TBE: mix the following: 108 g Tris base, 55 g boric acid, 9.3 g $Na_2EDTA$, and make to 1 L with distilled water. The pH of this solution is 8.3.

## Equipment

1. Electrophoresis power supply: to deliver 40 mA at 2 kV.
2. Gel electrophoresis apparatus: obtainable commercially from several suppliers; alternatively, "home-made" if workshop facilities are available.
3. Plasticard: Before use, Plasticard is washed thoroughly with detergent in order to remove material inhibitory for acrylamide polymerization.
4. Microcentrifuges. Ideally, two models required: a vertical rack model and an angle head model.
5. Repetitive dispenser: This is required to dispense 2 μL aliquots from a syringe to which adjustable pipet plastic tips can be fitted.
6. Gel dryer: This should be large enough to dry down two 20 × 40 cm sequencing gels.
7. 1.5 mL microcentrifuge tubes, with and without caps.

# Method

## *Preparation of the Template*

The M13 vectors described above contain a marker for identifying recombinants containing inserted DNA after ligation (*7*). Thus, when plated in the presence of an appropriate indicator, phage plaques containing inserted DNA are colorless, whereas those that contain no insert possess a blue perimeter. Templates are prepared, therefore, from colorless (or white) plaques.

1. A colony of JM101 is transferred from a minimal agar plate to 5-mL 2 × TY medium with a sterile inoculating loop and grown with vigorous shaking overnight at 37°C. A 1/100 dilution of this overnight culture is made into 2 × TY medium.
2. Each white plaque is then toothpicked into 1.5 mL of diluted JM101 in a glass culture tube and grown at 37°C for about 6 h with vigorous shaking. The culture is transferred to a 1.5 mL capped microfuge tube and centrifuged for 5 min in a microcentrifuge containing an angle-head rotor.
3. The supernatant is transferred carefully to another microfuge tube containing 150 μL of 20% PEG (6000), 2.5*M* NaCl. It is important not to transfer any of the bacterial pellet. If decanting, a portion of the supernatant should be left behind; alternatively, 1 mL can be removed with an adjustable pipet. The tubes are mixed by inversion and left at room temperature for at least 10 min before being centrifuged for 5 min in an angle head microcentrifuge.
4. The supernatant is aspirated, the tubes recentrifuged for 30 s, and the residual PEG is removed.
5. The pellet (which should be visible) is dissolved in 100 μL TE buffer by vortexing for 10 s. A 50 μL volume of buffer (0.1*M* Tris-HCl, pH 8.0) saturated phenol is added and the mixture vortexed for 10 s, left for 5 min, and revortexed for 10 s. The mixture is centrifuged for 2 min and the aqueous (upper) layer is removed using an adjustable pipet, taking care not to transfer any of the phenol layer.

6. One tenth volume of 2*M* NaAc is added, followed by 2.5 vol of EtOH. After mixing the tubes are left at −20°C overnight and the precipitate is recovered by centrifugation in a microcentrifuge for 10 min at room temperature.
7. The pellet is washed by adding 1 mL of 95% EtOH (without vortexing) and recentrifuging for 5 min. The pellet is dried in a desiccator under vacuum and dissolved in 30 μL TE buffer by vortexing for 10 s. The template solution is stored at −20°C.
   About 1–5 μg of single-stranded DNA is recovered by the procedure. The quality of the template can, if desired be checked by running 5 μL on a 1% agarose minigel and staining with ethidium bromide (*see* Chapter 7).

## Sequencing Reactions

1. In order to anneal the primer and template, 1 μL 17-mer primer (0.2 pmol/μL), 1 μL TM buffer, and 3 μL water are added *per clone* to a 1.5 mL microfuge tube. The 5 μL aliquots are dispensed into capped 1.5 mL Sarstedt tubes and 5 μL template DNA (about 800 ng) per tube are added. The tubes are centrifuged briefly to mix and placed in a dish of water prewarmed in a waterbath to 75°C. The dish is removed from the waterbath and allowed to cool on the bench for 15–30 min. The tubes are then centrifuged briefly to concentrate condensation to the bottom of the tube. The annealed samples should be used immediately and not stored.
2. The annealed primer-templates are used to initiate primed DNA synthesis. Into each of four siliconized 0.5 × 2 in. glass tubes, add an aliquot of 1 μL/clone of 1 mCi/mL [α-$^{32}$P]dATP (obtained in 50% EtOH at a specific activity of approx. 450 Ci/mmol). The label is dried down under vacuum and redissolved in 1 μL/clone of the dNTP mix and 1 μL/clone of one of the four ddNTPs. The working concentrations of ddNTP should be determined by assay; however, we use routinely 0.05 m*M* ddA, 0.15 m*M* ddG, 0.04 m*M* ddC and 0.4 m*M* ddT final concentration. Thus the ddNTP solutions

should be twice these concentrations since they are diluted twofold with the dNTP mix. The ddNTP–dNTP mix containing labled dATP is referred to as N°.

3. The subsequent sequence reactions are performed in 1.5 mL uncapped microfuge tubes in 10-hole microfuge racks. Solutions are dispensed with a repetitive dispenser. The four centrifuge racks are labeled A, G, C, and T, respectively. The arrangement of the tubes in these racks is summarized in Fig. 2. For dispensing solutions, the racks are placed vertically, such that the capless microfuge tubes (one per clone per rack) are horizontal. This prevents premature mixing of the sequencing solutions. Two μL of the primer-template solution is dispensed onto the side of a tube in each of the four racks. A 2 μL volume of the N° mix is then added: 2 μL of A° is added per tube to the rack labeled A, and so on. At this stage, the racks are centrifuged briefly (10 s) to mix the solutions prior to initiating primed synthesis with 2 μL of Klenow DNA polymerase I. The Klenow solution is prepared by diluting the stock solution of the Klenow fragment of DNA polymerase I into TE buffer to a final concentration of 0.125 U/μL and dispensing without delay. The tubes are centrifuged briefly to start the reaction ($t = 0$) which is allowed to proceed for 15 min at room temperature. During this time 2 μL of dNTP chase mix is dispensed onto the side of each tube. The tubes are centrifuged briefly at $t = 15$ min and left at room temperature for a further 15 min. Any chains not ending in ddNTPs are extended during the chase period to such a size that they will not enter the gel during the subsequent PAGE fractionation. The reaction is stopped at $t = 30$ min by addition of 4 μL of formamide dye mix. The racks are then placed in a boiling water bath for 2.5 min in order to denature the in vitro synthesized complementary strand from the parent template strand.

## *Acrylamide Gel Separation*
(*see also* Chapter 52).

After denaturation, the in vitro synthesized strands are separated according to their differing chain lengths by PAGE. Routinely, two 6% polyacrylamide gels are cast for

| A° | G° | C° | T° |
|---|---|---|---|
| 77μM dG | 5.4μM dG | 109μM dG | 109μM dG |
| 77μM dC | 109μM dC | 5.4μM dC | 109μM dC |
| 77μM dT | 109μM dT | 109μM dT | 5.4μM dT |
| 50μM ddA | 150μM ddG | 40μM ddC | 400μM ddT |
| 1μl/clone $^{32}$P-dA | 1μl/clone $^{32}$P-dA | 1μl/clone $^{32}$P-dA | 1μl/clone $^{32}$P-dA |
| 2μl/tube in A rack | 2μl/tube in G rack | 2μl/tube in C rack | 2μl/tube in T rack |

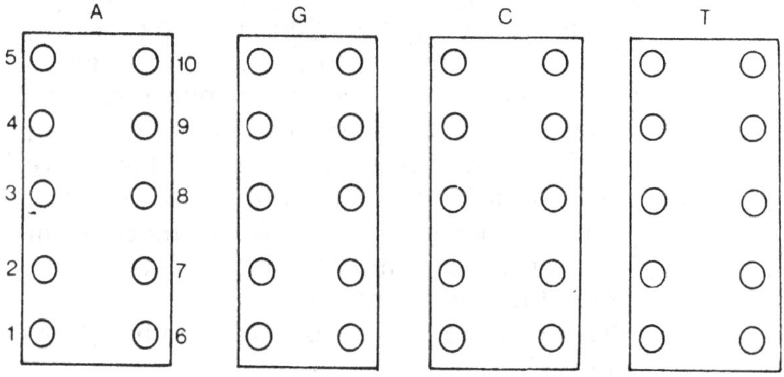

A G C T

5 10
4 9
3 8
2 7
1 6

Clone No. 1
1μl 17mer primer
1μl TM buffer
3μl $H_2O$
5μl template DNA

2μl into each of
A1, G1, C1, T1

Similarly for Clone No. 2 - Clone No. 10

Fig. 2. Composition of A°, G°, C°, and T°.

each sequencing run. One gel is electrophoresed for 2 h and the other for 4 h. This procedure yields about 300 bases of sequence.

1. Electrophoresis is performed on 0.3 mm thick 20 × 40 cm gels. For each gel, one notched and one unnotched glass plate are thoroughly cleaned by scrubbing with detergent, rinsing with hot water, finally with ethanol,

and then are dried. Gel plates *must* be very clean in order to prevent the formation of bubbles when pouring the gel. The notched glass plate is siliconized on the inner surface before assembling in order to facilitate the removal of this plate after electrophoresis.

2. The plates are assembled using two 0.5 × 40 cm strips of 0.3 mm Plasticard as spacers and the plates are clamped with foldback clips. Both sides and the bottom are sealed with PVC tape.
3. For each gel, 50 mL of 6% gel mix is used; polymerization is initiated by the addition of 40 μL of TEMED and 300 μL of 10% ammonium persulfate. After mixing, the gel solution is poured down one edge of the glass plates, held at an angle of about 45°C, using a 50 mL plastic syringe. During filling, the plates are lowered gradually to horizontal and the slot former is placed in position. The slot former is machined from 0.3 mm Plasticard; each comb contains 30 slots with teeth 5 mm deep, 3 mm wide, and 1.5 mm apart. The gel is left at least 30 min to polymerize, and can be left overnight if required.
4. Prior to assembling onto the gel apparatus, the PVC tape sealing the bottom of the glass plates is slit with a scalpel or, if preferred, removed completely. The slot former is then removed and the slots are flushed with TBE buffer to remove any unpolymerized acrylamide. After assembling, the reservoirs of the apparatus are filled with TBE and the gel is left for at least 1 h before loading.
5. The sequence reactions are loaded, after boiling, using a drawn out capillary tube. The tube should be drawn out such that it should fit easily into the slot and yet not be so narrow that filling and discharging become difficult. *Just prior to loading*, the slots are flushed again with TBE in order to remove any urea which has diffused from the gel. Failure to do this will make sample layering difficult. A volume of 1–2 μL is loaded; in between samples the capillary is rinsed in the bottom reservoir buffer.
6. It is essential to run the gel at about 50–60°C surface temperature in order to prevent any renaturation. This is achieved by running the gel at a constant 40 W (po-

tential about 1.2–1.5 kV; current about 28–32 mA). For the short run, the bromophenol blue is run to within an inch of the bottom of the gel (about 2 h), although this depends somewhat on the vector and the cloning site. The bromophenol blue front corresponds to the mobility of an oligonucleotide of about 20 bp. Since the primer is 17 bp, no insert sequence should be missed. The long run is normally electrophoresed for 4 h.

7. At the end of the run the PVC tape is removed and the notched plate is levered off with a scalpel blade. The removal of this plate should have been facilitated by its prior siliconization. The gel, still on the unnotched plate is placed in 10% acetic acid, 10% methanol, 80% water (2 L) for 15 min. This serves to fix the bands and to reduce the urea concentration. The glass plate and gel are then thoroughly drained of excess fluid and a sheet of Whatman 3 mm paper is placed on the gel. The gel sticks readily to the 3 mm paper when it is peeled off.
8. A layer of Saran Wrap is placed on the gel prior to drying on a gel dryer at 80°C for 30 min. The Saran Wrap is removed after drying and the gel is placed in direct contact with an X-ray film inside a film cassette and left overnight at room temperature before developing.

## Reading the Sequence

1. An example of a sequencing gel is shown in Fig. 3. The interpretation of the autoradiograph is subjective. Ideally, bands should be regularly spaced and of equal intensity. In practice, this is never so; regions of stable secondary structure give rise to aberrant band spacing, and band intensity can vary markedly as a function of the local sequence. Band compression can normally be resolved by sequencing the complementary strand or by running the gel at a higher temperature. Since band intensity is sequence specific, the following rules are useful in interpreting difficult regions of the gel:

   Upper C is always *more* intense than lower C.
   Upper G is often *more* intense than lower G,
      particularly when they are preceeded by a T.
   Upper A is often *less* intense than lower A.

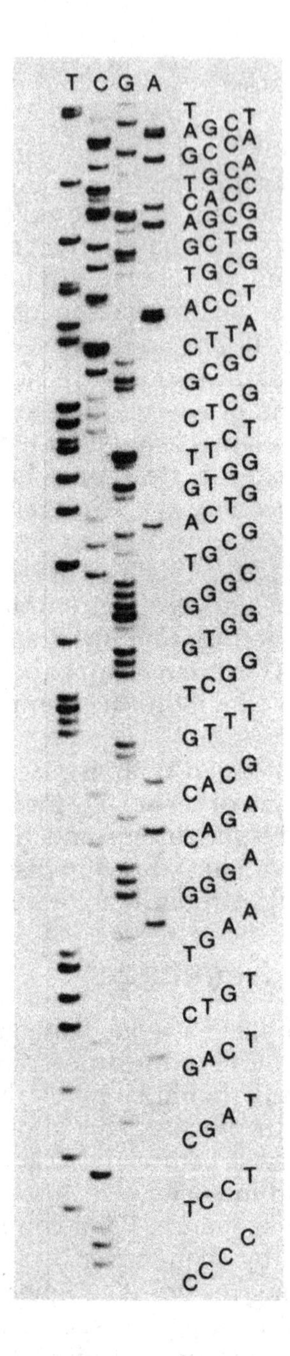


Fig. 3. A portion of the autoradiograph from a 6% polyacrylamide gel of a sequence reaction is shown, together with the sequence read from the autoradiograph. The sequence should be read in conjunction with the section in the text (p. 361) devoted to rules for reading DNA sequences.

2. It is essential to read spaces as well as bands. Since the spacing is generally regular, a double spacing might indicate the presence of an undetected residue in this position, for example, a single C band.
3. The beginning of a sequence will contain part of the polylinker sequence of the vector up to the cloning site. If the insert is sufficiently short the sequence will be read into the M13 vector on the 3′ side of the insert. It is important not to confuse this with insert sequence; the T track in this region (5′ doublet, six singlets, triplet 3′) is easily recognizable and should be used as a guide to the end of the cloned sequence.
4. Of the order of 300–350 bases can be read from the short and long gel runs. This comprises about 100–150 bases from the short run and as much again from the long run. Since eight clones can be loaded onto a gel, about 2000 bases of sequence can be obtained from two gel runs per day.
5. For the vast majority of sequencing projects, computer analysis of the data is essential. A variety of suitable programs have been written (*8*) and may already be available on computer facilities. All these programs, however, rely on accurate data. Before attempting to determine an unknown sequence, it is essential to determine one's competency by practicing on a template of known sequence, for example, the M13 vector. Only when these data are of high quality should the sequence project be started.
6. Finally, the ultimate aim should be to determine the sequence of *both* strands of the DNA.

# Notes

1. The quality of the DNA template preparation is crucial for successful dideoxy DNA sequencing. Contamination of the template with a variety of agents can influence severely the sequence reaction. In particular, the presence of PEG (used in the preparation of M13 phage particles) or sodium acetate (used in ethanol precipitation of the template) results in an inhibition of the DNA polymerase reaction. Contaminating host bacterial

DNA or RNA will also decrease the quality of the final autoradiograph, resulting in a general dark background.

2. Another common reason for failure is associated with the quality of the Klenow fragment of DNA polymerase I. The sequence reaction demands a highly active enzyme. Several suppliers assay their Klenow fragment preparations in a M13 sequencing reaction. Great care should, however, be taken upon storage; in particular, the stock solution should be diluted immediately prior to use in the sequence reaction and should never be allowed to warm up. An impure preparation of template can sometimes inhibit the enzyme activity (see above). If this is the case, there is no alternative but to retransfect competent host cells and prepare a new batch of template.
3. The sequence reactions can be stored at −20°C prior to PAGE fractionation. They should not, however, be frozen in the formamide/dye mix since this results in DNA breakdown and consequently an increased gel background. An increased background, and the appearance of artifactual bands, also occurs during prolonged storage at −20°C, presumably because of the accumulation of radiolysis products. Longer than overnight storage is not recommended.
4. The heat denaturation of the parent template and its complementary strand is also important. Failure to heat denature the reaction products fully prior to PAGE fractionation will result in intense radioactivity at the top of the gel and decreased intensity of bands lower down. Boiling of the samples for longer than 3 min will result in DNA breakdown and a higher background on the gel.
5. The PAGE fractionation step is generally trouble-free. Smearing of bands can result if the gel is run too hot, if the acrylamide solution is not deionized, or if too large a sample volume (>2 µL) is applied. Streaking can occur if urea crystals are present in the slots when the sample is loaded. For this reason, it is important to flush out the slots immediately prior to sample loading.

6. The sharpness and, therefore, the resolution of the bands increases when the gels are dried prior to autoradiography. The dried gels occasionally adhere to the film during exposure. This is a result of incomplete drying of the gel or of adsorption of water by residual urea still present in the gel after fixation. If a suitable gel dryer is not available, reasonable results can be obtained when the wet gel is covered with Saran wrap and exposed directly.
7. Recently, a modification has been introduced to both the gel system and the radioactive label (*9*). The effective separation range on a single gel has been increased by fractionating the reaction products on an increasing gradient of TBE buffer towards the bottom of the gel. The sharpness of the individual bands has been improved by the use of deoxyadenosine 5′-(α-[$^{35}$S]thio) triphosphate instead of [α-$^{32}$P]dATP as the radioactive label (*9*). This analog of dATP is a substrate for DNA polymerase I, and the short path length of the β-particles emitted by $^{35}$S results in very sharp band definition. These improvements increase the length of DNA sequence data that can be read from a polyacrylamide gel.
8. The problems associated with data analysis will depend upon the precise program used. It is worth reemphasizing, however, that too many uncertainties or mistakes in the sequence of the individual clones will complicate considerably the data analysis. It is futile, therefore, to commence a sequence project until the investigator has complete confidence in his sequencing technique.

## Acknowledgments

These protocols were developed in the laboratory of Dr B.G. Barrell, MRC Laboratory of Molecular Biology, Cambridge. I thank members of this laboratory, and particularly of Dr T.H. Rabbitts' laboratory, for teaching me dideoxy DNA sequencing.

## References

1. Sanger, F., Nicklen, S., and Coulson, A. R. (1977) DNA sequencing with chain terminating inhibitors. *Proc. Natl. Acad. Sci. USA* **74,** 5463–5467.
2. Sanger, F., and Coulson, A. R. (1978) The use of thin acrylamide gels for DNA sequencing. *FEBS. Lett.* **87,** 107–110.
3. Messing, J., and Vieira, J. (1982) A new pair of M13 vectors for selecting either DNA strand of double-digest restriction fragments. *Gene* **19,** 269–276.
4. Dunkworth, M. L., Gait, M. J., Goelet, P., Hong, G. F., Singh, M., and Titmas, R. C. (1981) Rapid synthesis of oligonucleotides VI. Efficient, mechanised synthesis of heptadecadeoxyribonucleotides by an improved solid phase phosphotriester route. *Nucleic Acid Res.* **9,** 1691–1706.
5. Deininger, P. L. (1983) Random subcloning of sonicated DNA: application to shotgun DNA sequence analysis. *Anal. Biochem.* **129,** 216–223.
6. Anderson, S. (1981) Shotgun DNA sequencing using cloned DNase 1-generated fragments. *Nucleic Acid Res.* **9,** 3015–3027.
7. Messing, J., Gronenborn, B., Muller-Hill, B., and Hofscheneider, P. H. (1977) Filamentous coliphage M13 as a cloning vehicle: Insertion of a Hind III fragment of the *lac* regulatory region in M13 replicative form *in vitro. Proc. Natl. Acad. Sci. USA* **74,** 3642–3646.
8. *Nucleic Acids Res.* (1982) Issue on applications of computers to research on nucleic acids. **10,** 1–456.
9. Biggin, M. D., Gibson, T. J., and Hong, G. F. (1983) Buffer gradient gels and $^{35}S$ label as an aid to rapid DNA sequence determination. *Proc. Natl. Acad. Sci. USA* **80,** 3963–3965.

# Index